T0337731

THE PEREGRINE RETURNS

WRITTEN BY Mary Hennen AND

ILLUSTRATED BY Peggy Macnamara

Foreword By John Bates · Photos By Stephanie Ware

THE PEREGRINE RETURNS

The Art and Architecture of an Urban Raptor Recovery

THE UNIVERSITY OF CHICAGO PRESS
PUBLISHED IN ASSOCIATION WITH THE FIELD MUSEUM

CHICAGO AND LONDON

The University of Chicago Press, Chicago 60637
The University of Chicago Press, Ltd., London

Published 2017
Printed in China

26 25 24 23 22 21 20 19 18 17 1 2 3 4 5

ISBN-13: 978-0-226-46542-5 (cloth)
ISBN-13: 978-0-226-46556-2 (e-book)
DOI: 10.7208/chicago/9780226465562.001.0001

Library of Congress Cataloging-in-Publication Data
Names: Hennen, Mary, author. | Macnamara, Peggy, illustrator.
Title: The peregrine returns : the art and architecture of an urban raptor recovery / written by Mary Hennen and illustrated by Peggy Macnamara ; foreword by John Bates ; photos by Stephanie Ware.
Description: Chicago ; London : The University of Chicago Press, 2017. |
"Published in Association with the Field Museum."
Identifiers: LCCN 2016034614| ISBN 9780226465425 (cloth : alk. paper) | ISBN 9780226465562 (e-book)
Subjects: LCSH: Peregrine falcon—Illinois. | Urban wildlife management—Illinois.
Classification: LCC QL696.F34 H46 2017 | DDC 598.9/609773—dc23 LC record available at https://lccn.loc.gov/2016034614

♾ This paper meets the requirements of ANSI/NISO Z39.48-1992 (Permanence of Paper).

CONTENTS

FOREWORD

In the Bird Department at the Field Museum, we give frequent tours of the collections, emphasizing the wonder associated with the feathered marvels we study, and we emphasize how important human interactions are for birds. People hear about the stories of extinction. Passenger Pigeons, Ivory-billed Woodpeckers (probably), Carolina Parakeets, and Bachman's Warblers are all North American birds that are now extinct because of humans, and there are others. But while we need to learn everything we can from such losses, we also can learn from our successes. Peregrine Falcons, the birds at the center of this book, tell an amazing story of the ability of humans to recognize an issue they have created for the animals around them and then actually do something about it successfully. In this case, we accomplished this by banning a pesticide that was strangling reproductive output in falcons and other predatory birds, and then bringing these spectacular aerial raptors back from the brink of extinction. The way this recovery has unfolded is the story told here through the eyes of someone who has lived it. Mary Hennen has spent much of her adult life thinking about these birds, while monitoring and promoting their recovery in Illinois. An important twist on this story is that Peregrines have, with help from humans, taken to a novel habitat; they now nest in the urban canyons of our Midwestern cities.

Through Peggy Macnamara's dynamic artwork and Mary's text, you will learn about and come to appreciate these magnificent birds, their new urban digs, and the value of promoting this conservation success because it is happening right outside the windows of human residents of some our biggest cities. You also will learn that human-wildlife interactions are never going to be easy or without issue in our urban environments. However, through Mary and her colleagues' efforts, Peregrines have caught the imagination of many of the human city dwellers with whom they now share territories. Along with Mary and Peggy, I hope that appreciation of these falcons and their stories is the tip of the iceberg, and that people can come to be more aware, accepting, and fascinated by the biodiversity around them and the need to study and monitor it. This is a growing responsibility we have as stewards, and Mary shows how exciting it can be (if you are not afraid of heights). Peregrine Falcons are a majestic species with which to initiate a connection with the rest of life outside your window.

JOHN BATES
Curator of Birds, the Field Museum

INTRODUCTIONS

Whether it is conducting research, educating future scientists, or preserving a single species, the Chicago Peregrine Program at the Field Museum plays a diverse role in advancing science and conservation. Together, scientists and citizens make a lasting impact on Peregrine Falcons and the environment in which they live.

Architecture expresses a city's past, present, and future. From neoclassical structures, to skyscrapers, to modern environmentally "green" buildings, Chicago is famous for its wide wealth of buildings, different in style and function.

Through the artwork of Peggy Macnamara and the voice of Mary Hennen, explore how Peregrine Falcons have recovered from a catastrophic decline and now call a man-made habitat home. In doing so, delve into the relationship between birds and the architecture of their urban environment.

Our ability to perceive quality in nature begins, as in art, with the pretty. It expands through successive stages of the beautiful to values as yet uncaptured by language.

ALDO LEOPOLD, *A SAND COUNTY ALMANAC*

SCIENTIST INTRODUCTION

The Field Museum houses an oological research collection of over 21,000 bird egg sets. Over the years scientists from around the world have used this collection in a wide variety of studies. One helped bring back a species from near extinction. That species was the Peregrine Falcon (*Falco peregrinus*).

By the 1960s, Peregrines in the United States had declined to less than 12 percent of their historic levels and were extirpated (wiped out regionally) from the Midwestern and Eastern United States. The few remaining nesting Peregrines were failing in attempts to breed. In an effort to understand the problem, scientists collected shell fragments from these nests, went to museums, and compared their specimens with eggs collected prior to the decline.

Scientists discovered DDT adversely affected breeding through the thinning of egg shells. Once the chemical was banned, the problem became how to bring Peregrines back to the wild. The Chicago Peregrine Program took on this challenge for Illinois. A few years after its start, I began with the program.

For nearly thirty years, I have monitored Illinois' Peregrines. I have seen the population grow from none, to a single pair, to nearly thirty today. Peggy's artwork illustrates my journey as well as the Peregrines, a species climbing back to a healthy breeding population.

ARTIST INTRODUCTION

John Ruskin, in his 1857 book The Elements of Drawings, refers to "the advantage of perfect fellowship, discipline and contentment." This has been my experience painting the Peregrine story. The thirty-year saga of the cliff-dwelling Peregrines finding their way to Chicago, a city renowned for its architecture, afforded me endless composition opportunities to blend the order of architecture and the freedom in nature. The Peregrine, the fastest bird on earth, can be seen in a variety of gestures like diving, soaring, protecting, and bowing, all of which are balanced by the structure and stability of architecture. In adapting to its new environment, the Peregrine adapted to what was available. The strength of drawing architecture is its clean and decisive lines. The diagonal is a valuable tool, as a compositional device to bring the viewer into the picture. The Peregrine, in almost any pose, can move the eye around, and finally the strong verticals and horizontals of a building element can ground the whole composition. The bird and its buildings have the advantage of perfect fellowship.

So the subject of a Peregrine comeback offered artistic opportunities as well as purpose. Much of art today is about . . . I hesitate here because I am not sure how to honestly finish my statement. Rather than surprise, stun, or offend, I choose to paint about nature, its magic and wonder. Rather than concentrate on destruction, I choose to highlight what the scientist in conjunction with the citizen scientist can accomplish. The Peregrine story gives us hope and encouragement. I want to celebrate and participate in turning things around.

The Field Museum community is a beehive of activity exploring, explaining, and recording what is happening on our planet. Each member of the community contributes something unique. I hope to touch on all the good that can come out of putting our heads together. The scientist, artist, designer, writer, naturalist, and many others all can come together at an institution like the Field Museum. Here this diverse group can create positive change. In this book we have added to the Peregrine story and touched on some of the ways science can contribute to our understanding, change our actions, and improve our planet. In the words of John Ruskin, these stories are "the advantage of perfect fellowship, discipline and contentment."

CHAPTER ONE

Decline of the Peregrines

SCIENTIST NOTE: *By the time I came on the scene, Peregrines had already been extirpated from Illinois. I was not specifically a birder growing up but I loved wildlife, especially observing behavior. During annual vacations in northern Wisconsin, you might find me with my grandma looking for White-tailed Deer under the trees. I loved how they slept under the pines. Along with my siblings and cousins, we collected toads, chased crayfish, and ran from bats flying at night. At night our parents might take us to watch Black Bears feeding at the dump. Nature was our playground. Still, for me growing up, there would never have been an opportunity to see Peregrines in the wild simply because they were not there anymore.*

Historically, an estimated 350–400 pairs of Peregrine Falcons once nested in the Midwestern and Eastern United States. By the 1960s, the species had been extirpated and few were seen during migration. The primary cause for the decline was determined to be an environmental buildup of organochlorines, in particular DDE, a by-product of DDT.

DDT was one of the first chemicals widely used as a pesticide. Sprayed on croplands to control insect populations, large amounts were accumulated in the bodies of Peregrines, a predator at the top of the food chain.

These accumulated chemicals caused addling of eggs, abnormal reproductive behavior in adults, and thinning of shells, which led to egg breakage. As a result, female Peregrines were laying eggs that were 10–20 percent thinner than normal, leading to nest failure across the species' range.

The last recorded breeding by Peregrines in Illinois before they disappeared from the state occurred in 1951.

Peregrine Falcon
Falco peregrinus
Order: Falconiformes
Family: Falconidae

Peregrine Falcons are a mid-sized raptor found on every continent except Antarctica. Their name comes from the Latin word *peregrinus*, which means "to wander." The species is known as the fastest flying bird, being able to dive at speeds of over 200 mph. In North America, Peregrines are roughly the size of crows. Adults are slate blue-gray above with barred underparts, while juveniles are a dark brown on their back and have heavily vertical streaking on the front. All ages have thick sideburns, called malar stripes.

MUSEUM EGG COLLECTION

By comparing modern-day eggs to those in collections made 100+ years ago such as those at the Field Museum, scientists can study various issues such as habitat change, effects of pesticides, or how climate change has impacted clutches.

EGG SHELL THINNING

DDE (a by-product of DDT) inhibits calcium production in female Peregrines, resulting in a reduced thickness of egg shells. The weight of incubating adults crushed the eggs before they could hatch. Scientists discovered this by comparing shell fragments from failed nests during the height of the decline to historic eggs from museum collections.

Cliff-dwelling Peregrines

Peregrines disappeared from Illinois' cliffs by the early 1950s. By the 1970s they were extirpated from the Midwestern and Eastern U.S. and declining worldwide. Efforts were made to restore a species that historically was present, but nearly wiped out due to human influence.

- Late 1940s – DDT introduced as a synthetic insecticide.
- 1972: U.S. bans the use of DDT.
- 1973: Peregrine Falcons were placed on the U.S. Endangered Species List.
- 1975: First releases (hacking) of Peregrines in the Eastern U.S.
- 1982: First releases (hacking) of Peregrines in the Midwest.
- 1985: Chicago develops plan to help reintroduce Peregrines into Illinois (implemented in 1986).
- 1999: Peregrine Falcon removed from U.S. Endangered & Threatened Species List.
- 2015: Peregrine removed from Illinois' Endangered & Threatened Species List.

CHAPTER TWO

Effects of DDT

SCIENTIST NOTE: *In September 2015, Common Nighthawks* (Chordeiles minor) *were observed migrating over Chicago in massive numbers (8800+). I didn't see nearly that many but even seeing the few I did both surprised me and reminded me of my childhood.*

There is a streetlight at the end of the driveway of the house where I grew up. I happen to be quite fond of that light. Not just from the amusement of watching people try to back out of our driveway when cars were parked on the street. (Picture dented pole and dented fenders.) But because of a fascination with what flew around it at night.

My father had a rule that you had to be home by the time the streetlight was on. You could still play in the yard, but you had to be home. As a result, a lot of nightly activities took place on our front lawn. Catching fireflies or just sitting on the driveway watching bats come to feed on moths attracted to the streetlight were favorite pastimes. Nighthawks flew by as well.

I now realize how precious those memories are. You rarely hear the buzz of a nighthawk anymore, as populations declined by an estimated 60 percent during 1966–2010 according to scientists at Cornell. Populations have been reduced as a result of a number of factors including mosquito abatement and loss of gravel rooftops. Perhaps the species is doing better than I thought considering recent migration numbers.

One key to the recovery of Peregrines was discontinuation of DDT use. This benefited other animals as well, since the negative effects of DDT reached far beyond a single species.

Birds such as Bald Eagles and Brown Pelicans were impacted in the same manner (thinning of egg shells) as Peregrines. Other species had their prey affected. For example, with an absence of insects, the Carlsbad Cavern population of Mexican Free-tailed Bats dramatically declined from numbers reaching 8.7 million down to 200,000.

The impact of DDT was felt even more strongly by species already in decline. Fireflies (lightning bugs) were already in trouble as light pollution from artificial lights made it difficult to see the flashes used as breeding signals. Urban development was destroying their habitat. On top of these factors, DDT eliminating their prey made firefly populations sparse.

A ban on DDT was put in place in the U.S. in 1972. It wasn't until 2001 under the Stockholm Convention that a worldwide ban on its agricultural use was enacted. To date, a limited use of the pesticide remains as a vector control for malaria-causing mosquitos.

Mexican
Freetailed
Bat
2014

Bald Eagle
Haliaeetus leucocephalus
Order: Accipitriformes
Family: Accipitridae

The Bald Eagle has been the U.S. national symbol since 1782. During the 1970s the U.S. population declined by over 70 percent. Like Peregrines, eagle populations have surged due to recovery efforts. Bald Eagles can live up to thirty-five years in the wild. It takes five years for an eagle to get its white head.

Brown Pelican
Pelecanus occidentalis
Order: Pelecaniformes
Family: Pelecanidae

Brown Pelicans are stocky seabirds. Their oversized bill has a stretchable throat pouch. They feed by plunge diving into water, using the force of the impact to stun small fish before scooping them up. Their numbers were greatly reduced in the 1970s, but they are now common again along the Atlantic, Pacific, and Gulf Coasts.

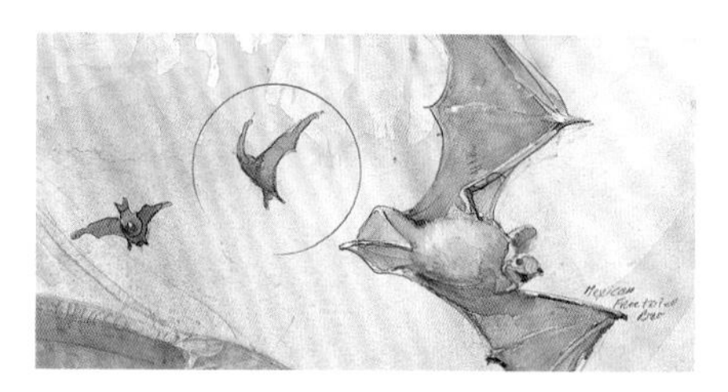

Mexican Free-tailed Bat
Tadarida brasiliensis
Order: Chiroptera
Family: Molossidae

Mexican Free-tailed Bats are found in the Western United States. They migrate to Central America and Mexico during winter. Some roosts, such as those at Carlsbad Caverns, New Mexico, are estimated to contain millions of bats, which eat over 250 tons of insects every night.

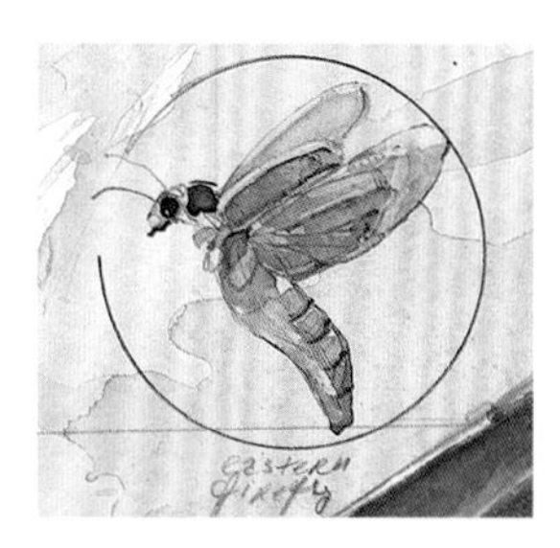

Eastern Firefly
Photinus pyralis
Order: Coleoptera
Family: Lampyridae

The Eastern Firefly is just one of two thousand species of lightning bugs. Fireflies get their name from the ability to light their lower abdomen in order to attract a mate. The Eastern Firefly is also called the Big Dipper Firefly, which refers to the J-shaped flight of the male, who lights up on the upswing.

CHAPTER THREE

Reintroduction

SCIENTIST NOTE: *My work with Peregrines began as the main hack site attendant during Illinois' fourth year of release (1989). I lived at Illinois Beach State Park on Lake Michigan for four days each week. My day started at sunrise, when I left camp, climbed the hack tower, and fed the birds. It ended at dusk, when darkness fell and the birds were no longer visible. It was an amazing experience. Animal behavior is fascinating and I was spending more than forty hours a week watching young Peregrines.*

One morning I watched for hours as Peregrines played in the surf. One would lie down on shore and let the waves wash over him. As a rolling action tumbled the falcon, it would stand, run up the beach returning to lie down again. Others joined in. Immature falcons acting like a bunch of kids frolicking in the surf. Being enveloped in the delight shown by the Peregrines, I was able to put aside any worry over the species to simply enjoy the show.

How were scientists able to bring back a species no longer found in the wild? It was not enough just to ban DDT (which occurred in 1972) or place Peregrine Falcons on the Endangered Species List (1973). While those measures would help protect the few remaining Peregrines, it did nothing to restore them to areas where there were none.

Scientists with the Peregrine Fund (then based at Cornell University) developed a program of captive breeding and release, called hacking. The first two releases of Peregrines in the Midwest (near the Mississippi River) in 1978 and 1979 involved birds bred at the Cornell lab. Beyond that, with the exception of perhaps one more from Cornell, all of the birds released in the Midwest came from falconers turned breeders.

Illinois used hacking methods when the Chicago Peregrine Program joined the effort to reestablish Peregrines in the Midwest. Our goal was to have three nesting pairs in Illinois by 1990, the last year of release.

Chapter 3
Reintroduction.

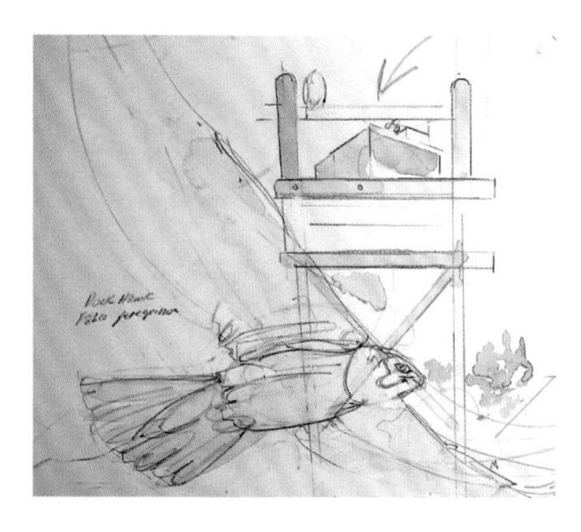

The Chicago Peregrine Program began in 1985 as a cooperative effort of the Chicago Academy of Sciences, Lincoln Park Zoo, Illinois Department of Conservation, and Illinois Audubon Society. Its goal was to help restore Peregrines to Illinois and the Midwest. From 1986 to 1990, the Chicago Peregrine Program released a total of forty-six captive-bred Peregrines from four different hack sites.

In order to have young Peregrines to release into the wild, scientists first had to find a source of birds. They turned to falconers. By breeding their birds in captivity, falconers could supply scientists with fertile eggs and/or chicks. Eggs were hatched in incubators and the young chicks (called eyasses) were carefully raised to ensure the birds would stay wild and not imprint on people.

The Raptor Center in Minnesota coordinated the transfer of young Peregrines from breeders to the Chicago Peregrine Program. The young birds sent were approximately thirty-five days old, about a week younger than fledging age. They were placed in a specially designed box, called a hack box, located at a chosen release site.

One side of the hack box is a sliding barred door open to the air. When it is closed, young falcons can still observe their surroundings but not venture out. A "delayed feeder" allows hack attendants to feed the birds without being seen. Prior to release, young falcons are placed inside a back room with a trap door. While the birds are hidden, a two- to three-day supply of food is placed in the box and the front barred door is removed. Attendants back away from the hack box and the trap door is sprung. They are free.

Within the next couple of days the immature falcons will take their first flight. Care is taken to minimize the presence of humans so that birds are not prematurely startled into

fledging. Over the following weeks, the Peregrines learn, on their own, to fly and hunt. As they begin catching their own prey, the young falcons return to the hack site less frequently. Exploratory flights become longer and longer, with the Peregrines eventually leaving the site entirely to find their way in the wild.

The goal of the Chicago Peregrine Program was to help reestablish Peregrines on a regional basis with a hope that some birds would breed in Illinois. The success of this plan can be seen in the dispersal of Peregrines across the Midwest and their subsequent breeding. Not only have Peregrines released or born in Illinois moved to neighboring states but their birds have come here. Illinois' breeding adult Peregrines have originated from hack sites and eyries located in Illinois, Indiana, Minnesota, Iowa, Missouri, Ohio, Nebraska, Wisconsin, and Canada.

CHAPTER FOUR

Peregrine Life in the City

SCIENTIST AND ARTIST NOTE: As Peregrines spread throughout Chicago, it provides us an opportunity to learn more about our city. We learn not only how to navigate the streets, but we can explore the skyscrapers and structures that form this urban environment that Peregrines use. For the scientist, details of nest site selections can be gathered by studying where the Peregrines lay their eggs. The artist is drawn to the lines of the structural makeup of nest ledges and parallel features between sites.

Whether artist or scientist, we both enjoy the rich history and diverse architecture found in Peregrine site locations. Of equal enjoyment is discovering avian representation on some of the buildings as we walk to our Peregrine sites.

By far the most frequent species seen are owls. Near to the Metropolitan Correctional Center (MCC) site, the roof of the Harold Washington Library features a number of owls, the symbol of wisdom. Close to our Millennium Park site, sitting atop the peak of the University Club Building, is another perched owl.

As we look after our city birds, their architectural representations look over us.

Naturally a cliff-dwelling species, released Peregrines found a home in cities. Building ledges became pseudo-cliffs. Peregrines found ample prey and little nesting competition in their urban setting. In 1987, the first pair to nest in Illinois after the decline made its home on a building ledge near Chicago's famed Sears Tower (now called the Willis Tower).

Peregrines reside in Illinois year-round. Courtship begins in late winter, with the first eggs laid by late March or early April. By midsummer young Peregrines are fledging from their nests. While immature Peregrines will migrate out of the area in the fall, the majority of Illinois' adults stay through the winter.

Many people who work or live in city skyscrapers have a bird's-eye view into the life of our Peregrines. These individuals have been crucial to the Chicago Peregrine Program's ability to monitor the falcons. Those who are not fortunate enough to be close to where our Peregrines hang out can observe the nesting season through web cameras placed at some of the sites.

Typically a cliff dwelling species, Peregrines use city ledges as a suitable substitute. Building no formal nest structure, they lay their eggs in a "scrape," a small depression made in the substrate on a ledge. Adult Peregrines may make multiple scrapes as part of courtship. Ultimately the female will select where she will lay her eggs. One egg is laid every twenty-four to forty-eight hours until the clutch is complete. Clutch size can range from one to five eggs though on average three to four eggs is typical. In Chicago the first eggs are generally seen in mid to late March.

Peregrines delay incubation until the clutch is nearly complete, which ensures the young will all hatch at around the same time. Incubation lasts thirty to thirty-two days, so in Chicago most eggs begin hatching around Mother's Day. While male Peregrines assist with incubation, females are primarily responsible for brooding their young. This involves covering the eyasses by tucking them underneath a wing to either protect the chick from the cold or shade them from the sun.

Peregrine Chicks develop at an astonishing rate.

- A newly hatched falcon chick weighs about one and a half ounces (35–40 grams) and is virtually blind.
- Two weeks: Eyasses are four times their hatch size. The adult female Peregrine is no longer able to brood all the chicks due to their larger size.
- 21–24 days: Flight feathers and body contour feathers are poking through the down. The young birds are active, moving out of the nest scrape and exploring around the nest ledge. This is the perfect age for banding if the nest site is accessible to humans.
- 35–40 days: Most of the down is gone except for a few head tufts. The young birds exercise their wings by energetically flapping them.
- Six weeks: Eyasses take their first flights away from the nest. The first flight may be a glide to another ledge or ground level.

21–24 days old

35 days old

CHAPTER FIVE

Behaviors

SCIENTIST NOTE: *What has struck me over the years is how much the behavior of immature Peregrines, at times, resembles that of kids at play. Sibling rivalry is displayed when fighting over food the adult brought to the nest. New fledglings will approach each other in midair, locking talons and then flying off.*

*One day at a hack site, a Great Blue Heron (*Ardea herodias*) came slowly flying by. One of the young Peregrines went flying after it, matching the heron's speed. The falcon then picked up speed, flew low and up from behind with feet forward. Reaching out with its talons, the falcon swatted at the wings of the heron. As the heron tumbled and slowly righted itself, the Peregrine flew in a wide circle away and back behind the heron. This behavior repeated itself a half dozen times with the falcon tumbling the heron and the heron straightening itself out until the falcon had enough of that play and returned to the hack tower.*

*The next year at another hack site, one of the Peregrines would fly low over the head of a Mallard (*Anas platyrhynchos*) on a pond, forcing the duck to dive in reaction to a perceived threat. It would circle while the duck was underwater and again dunk the duck when it came back up. The Peregrine did this over and over again until it tired of the play.*

The scientist in me notes how the falcons were learning hunting techniques, but part of me just smiles at a child's mischievous behavior.

Each movement a bird makes is some type of behavior. In fact, this entire book classifies Peregrine actions into some kind of category. Nesting, feeding, breeding, flying . . . . This chapter highlights a few select behaviors but by no means attempts to list every one.

While general descriptions of behavior can be made, it does not follow that all birds will react to a given set of circumstances in exactly the same way. For instance, while most adult Peregrines will protect their young, the level of aggressiveness in defense can vary greatly among individuals.

COURTSHIP BEHAVIORS: BOWING

Courtship behavior includes ledge displays, individual (solo) or mutual. Bowing can take many forms, such as: Horizontal Head-Low Bow, Extreme Head-Low Bow, Vertical Head-Low Bow, and Agonistic Head-Low Bow. The pictured mutual Horizontal Head-Low Bow is often done as pairs vocalize a Creaking Call.

EGG LAYING

Female Peregrines lay an egg every twenty-four to forty-eight hours until the clutch is complete. As an egg moves down the female's oviduct it presses against glands that produce colored pigments. Peregrine egg colors range from pale creamy to a dark rusty brown.

Marking patterns develop on eggs when they reach the pigment glands. If an egg is stationary when coming in contact with the glands, spots develop. If the egg was in motion, it gets streaks.

PREENING

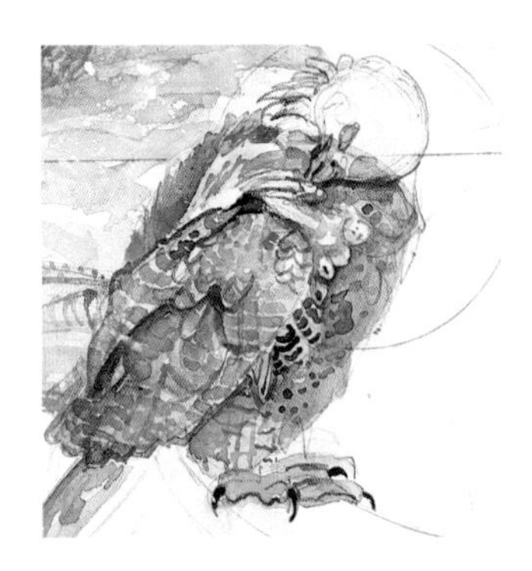

Like all birds, Peregrines preen their feathers to keep them in top condition. During preening, the bird may remove dust particles or parasites while realigning feathers into the best position in relation to adjacent feathers.

Using its bill, a bird draws oil from the uropygial gland at the base of the tail to spread to each feather, keeping it flexible and waterproof. Peregrines also erect all their feathers at once and shake them. This behavior is called a "rouse."

BATHING

It is widely accepted that birds take baths to help preserve their feathers. In Chicago, you might see a Peregrine on the beach hunting waterfowl or it may just be taking a bath alongside them.

Peregrine feathers get replaced (molted) once a year. In the interim, bathing and preening help to keep them in top condition. After all, feathers are their lifeline for not only flight but also keeping the bird waterproof or insulating them from cold temperatures.

BROODING

While male Peregrines are equal partners in incubating, brooding the chicks falls primarily to the females. For the first few days of life, young chicks are not able to keep warm on their own so the mother falcon tucks the small birds under her wings or breast feathers. Eventually the chicks are large enough that they no longer fit under the adult. At this stage they are able to thermoregulate (maintain their body temperatures) on their own.

FEEDING

Peregrines do not regurgitate food but instead present small morsels of meat to the hatchlings. Chicks learn to whine and gape their mouths open at feeding time.

AERIAL FOOD TRANSFERS

An aerial food transfer may involve passing prey from one bird's talons to another's while in flight. An immature fledgling Peregrine may take food from a parent, or an adult male Peregrine may pass along something to the female during courtship. This flight skill (flipping upside down to lock talons) is also useful in territorial fights.

CHAPTER SIX

Nest Site Selection

ARTIST NOTE: *Here the story gave me an opportunity to slip in a figure, the statue of St. Michael. I did this piece early on and knew I was hooked on the subject matter. Usually the artist finds he or she puts in the primary image and then is left questioning "what to do with the background?" Here the two elements, figure and ground, are equals in interest, form, and color. I was able to move from primary subject to background with equal affection. This constitutes good composition.*

The urban environment has provided some interesting nesting locations for Peregrine Falcons. Chicago's skyline is a rich heritage of architectural styles and buildings that were never designed with the avian world in mind are now called home by these falcons.

After being extirpated from Illinois, the first Peregrines to breed in recent history chose a ledge on a downtown Chicago office building as their nest site. As the population grew, falcons selected sites from historic landmarks to cultural meeting places and everything in between. Though the architectural styles of nest buildings vary greatly, specific aspects of a Peregrine nest ledge follow a general pattern.

DISTRICT
7
5

Wacker: Illinois' first postdecline nest site

Site Information: 1987–2015

Total eggs laid: 129

Eggs hatched: 77

Young fledged: 69

Rahn at Wacker site

Photo: S. Ware, 15 May 2013

Rahn

Band identification: b/g 01/A, 1807-77766

Natal site and year: Sheboygan, Wisconsin, 2001

Breeding location: Chicago, Wacker site, 2006–13

Rahn was the fourth female Peregrine to nest at the Wacker site. In 2011, she laid three eggs before her mate (unidentified at the time) was lost in a territorial fight. Rahn abandoned those initial eggs and laid two more clutches (four, then seven unfertilized eggs) while on her own. The following year Rahn, still alone, laid nine eggs total. That comes to twenty-three eggs in a two-year period. Rahn found a new mate the following year.

The first location where Illinois' Peregrines attempted to nest after the decline was the Wacker site in Chicago. This eyrie has been occupied by either a pair or a single bird annually for the past twenty-nine years (1987–2015). People can be under a mistaken impression that Peregrines chose nest locations by selecting the highest spot in their territory. This coincides with the idea Peregrines must be nesting on the roof. Both assumptions are usually wrong.

Nest site selection has little to do with height and more to do with a ledge's architectural features. Historically, Peregrines nested on the side of cliffs. City blocks provide a pseudo-cliff, and most urban Peregrines use ledges on the side of buildings. A particular ledge is chosen if it has certain desirable features. Peregrines need to be able to see all around them when lying down to incubate eggs. If the ledge has walls and the walls are too high, then it is not functional for the birds. Preferably the ledge is protected from prevailing winds. The Wacker ledge design was perfect for Peregrines, as it contained all the features the falcons look for in a nest site.

Ballistic is aptly named. She is one feisty, determined Peregrine. Originally from Ohio, Ballistic was first sighted in Chicago in 2006, the year after she fledged. Her first selection for a nesting site had all the desirable ledge features but was destined to fail. Ballistic chose a gutter, which flooded the eggs after the first rain. Ballistic then found her way to St. Michael's Church, taking the territory over from its previous female, Kelliwatt. When renovations to the church prevented use of her favored spot for a nest scrape, Ballistic headed for the Oak Street Beach neighborhood.

Since then, each year Ballistic has been forced to search for a new site owing to her selection of flower pots on privately owned condo balconies. Since Peregrines are protected, people are not allowed to harm either the adult falcons or their eggs. Once the young Peregrines have fledged, it is perfectly legal for building owners to prevent any future use of their site (such as removal of a flower pot) as long as no birds are harmed in the process.

Ballistic at gutter location
Photo: S. Ware, 21 November 2006, Wrigleyville (Belmont/Addison) site

Ballistic
Band identification: b/g 69/C, 1687-01651
Natal site and year: Bohn Building, Cleveland, Ohio, 2005
Breeding locations and years: Belmont/Addison, Chicago, 2006; St. Michael's Church, Chicago, 2007–11; Oak Street sites (three different buildings, 2012–14

St. Michael's Church nest site

Site Information: 2009–15
Total eggs laid: 13+?
Eggs hatched: 13+?
Young fledged: 13

Abandoned buildings, McKinley site

Site Information: 2009–15

Total eggs laid: 13+?

Eggs hatched: 13+?

Young fledged: 13

Immature peregrines, ready to fledge

Photo: S. Ware, 14 June 2014, McKinley site

While Peregrines are not cavity nesters, they will take advantage of a structure that imitates a protective cliff overhang. Such is the case with the McKinley site. An abandoned clock tower that is part of the historic buildings in the Central Manufacturing District on Chicago's south side has been home to a pair of Peregrines for six years (2009–14). The nesting falcons used a ledge just inside one of the missing numbers in the clock.

Because this nest is inaccessible and without a view inside, we cannot know an exact count on eggs laid. We can determine the minimum number by counting the greatest number of immature Peregrines visible at one time. A downside to an inaccessible nest is that we can't band the chicks. Without bands we have no means to study individual longevity or dispersal. Still, the ultimate goal is to have Peregrines successfully breeding. Getting to band chicks is a luxury. In 2015, only a single adult was present at the McKinley site.

CHAPTER SEVEN

Nest Fidelity

ARTIST NOTE: *I have found that the best way to lay in a strong composition is to delay commitment. You begin by softly suggesting where the various elements might lie and how the viewer's eye will move from one to the next. I know color plays a strong part in the composition but for now you decide only in the present. Adjustments, and there will be many, come as the piece develops. As in the case of Peregrine nest locations, there are no absolute rules.*

The old adage "there is an exception to every rule" holds true when discussing Peregrine behavior. Look at selecting nest locations. It is generally accepted that Peregrines exhibit strong nest site fidelity. This means they are faithful to their nest sites and will use the same location year after year, choosing a new site only if they fail to breed successfully or if the location becomes unusable.

So what does it mean when Peregrines don't follow that rule?

For example, Illinois' original pair of river Peregrines relocated their nest annually even after successfully fledging young at each site. More typical are pairs that use a single nest ledge for years.

We may never know why that one pair of Peregrines kept moving around. The cause for nest relocation may have resulted from issues with the site itself that were unseen by us. It may have been driven by individual choice. As we continue to follow the Peregrine population and its breeding, perhaps we will discover the answer.

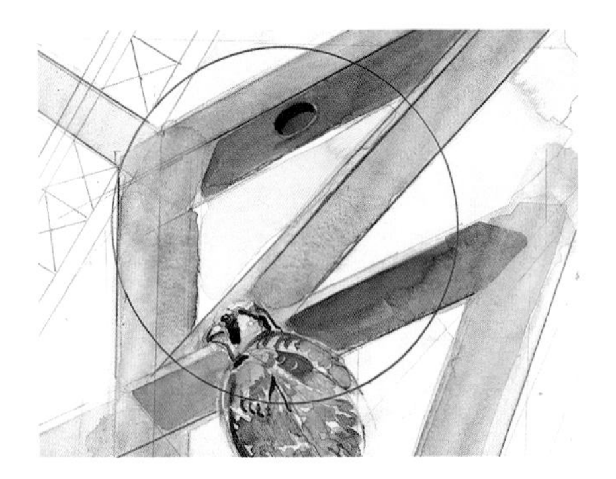

Top: Suboptimal habitats: bridges

Bottom: Exposed nest location, South Loop
Photo: S. Ware, 21 April 2010, South Loop site

Peregrine egg clutch
Photo: M. Hennen, 21 April 2015, Millennium Park site

Site Information: 2010–15
Total eggs laid: 22
Eggs hatched: 19
Young fledged: 17

Millennium Park building, opened 1929
Architect: Samuel N. Crowen & Associates

Peregrines don't always pick the best of spots to nest. While a gutter may have all the features the birds are looking for, ultimately the nest will wash out after heavy rains. Nesting in flower pots on balconies creates its own set of complications. Bridges may be appealing but fledglings are at risk of drowning on their first flights. Failure to breed successfully is a common cause of site abandonment.

Chicago's South Loop Peregrine site is an excellent example of suboptimal habitat. In other words, the location is acceptable but lacks features found in prime habitat. This site has a flat open main roof with a small central structure in the middle. The central structure is enough to make the roof the equivalent of an extremely wide ledge, but the eggs, chicks, and adults are exposed to extreme weather and vulnerable to predation. The South Loop site was improved by installation of a nest box, thereby increasing the chance for a successful breeding attempt.

With high population numbers, intense competition for nesting sites ensues. Some Peregrines are forced to use lower-quality sites.

Peregrines are considered to be monogamous and faithful to their nest site. The pair at Millennium Park have remained together and breeding on the same nest ledge for years. Occasionally a young adult Peregrine may spend just a year in a territory before relocating to a new site where it will breed with a new partner for years. This occurs despite the fact that the old partner and site are doing fine.

For example, in 1999, a female named Zoom (b/r *4/H) bred with a male named Tracy (b/r *P/M) at a church in Evanston, Illinois. They fledged one young. Though successful at breeding, and with no alterations to the church nest site and nothing wrong with her previous mate, Zoom choose to relocate to the Uptown Theater. Zoom remained at the theater for the rest of her years as a breeder. While Tracy stayed at the church for another year, he ultimately returned to his natal site, where he bred successfully for a number of years.

From 1994 to 2002, a pair that held a territory along the Chicago River used several locations regardless of breeding success. The pair, called the River Birds, used five buildings, including the Jewelers Building, over an eight-year period. It is unknown why this pair (Fast Eddy and Oog) continued to move even after successfully fledging young and no changes had occurred to their nest ledges.

One difficulty in monitoring Peregrines is that there are times when you just do not get all the information you would like. When you have a pair that annually relocates anywhere within two square miles in a congested urban location like Chicago, you can miss finding the pair until after the young have fledged. This was the case in 2005 when the River Pair fledged two young. They were discovered only when one of them, Autumn Hope, became injured. After a few months in rehab, she was banded and released in Lockport, Illinois, some thirty-five miles southwest of Chicago. Five years later, Autumn returned and was identified as the adult female in her natal territory. Unlike the original River Pair, Autumn favored using one building, the London Guarantee, the first location used by her parents in 1994.

River Peregrines at Jewelers Building

Autumn Hope: daughter of River Pair

Photo: S. Ware, 25 May 2010, London Guarantee location

Autumn Hope

Band identification: b/g 22/C, 1687-01960

Natal site and year: Near the Chicago River (exact location unknown), Chicago, 2005

Breeding locations and years: Chicago River Sites, 2010–15

CHAPTER EIGHT

Flight

ARTIST NOTE: *Many of my students are animators and this fact inspired me to attempt to paint movement. I have borrowed from the animators the use of successive poses and lightness of touch. The bird ceases to move when every detail is rendered. The fastest bird on earth has given me endless opportunities for slow study.*

Monitoring Peregrines has its own challenges, but it also affords an opportunity to witness spectacular behavioral displays. Many of these take place in the air.

Flight is part of most aspects of a Peregrine's life. It functions as part of courtship where the falcons may exhibit aerobatic displays of spiraling and soaring or exchange food on the fly. Speed and agility abound when, in defense of the territory, the Peregrine must chase a third individual away from the nest. Who can forget the first time they witnessed a Peregrine diving after prey? With speeds in excess of 200 mph, they streak after prey in an amazing demonstration of hunting skill.

From Leonardo da Vinci to the Wright Brothers to every child who has made a paper airplane, flight has been fascinating people for generations. If you are lucky, sometime in your life you will be able to watch a Peregrine soar.

The human bird shall take his first flight, filling the world with amazement, all writings with his fame, and bringing eternal glory to the nest whence he sprang.

LEONARDO DA VINCI

DEFENDING THE TERRITORY

As Illinois' Peregrine population has increased, it follows that the number of skirmishes between resident Peregrines and intruders has also risen. A strange Peregrine may be greeted by cacking calls before being chased out of the territory. In more aggressive encounters, one or both of the adults in a pair will attack the intruder. The attacker spirals up in flight to gain height in order to stoop down. Meanwhile the intruder has flipped over in flight onto its back to present its talons. Actual blows may be struck and occassionally the two adults may stay with locked talons as they fall to the ground.

STOOP

Peregrines reach the speeds they are famous for (over 200 mph) when diving after prey. This dive is called a stoop. Adaptations for these stoops include pointed wings but also a nictitating membrane (a transparent "third eyelid") that protects the eyes from dust and debris in the air. A special cone or baffle in the nasal cavity regulates the amount and speed of the airflow through the opening, making breathing easier.

FLEDGING

A Peregrine's very first flight (called fledging) is generally a glide down from the nest ledge to a level in line with or below the nest site. Occasionally, a fledgling will go all the way to the ground. We call this grounding. In the city, if a fledgling grounds, it needs to be retrieved quickly due to the high levels of activity on most city streets. We pick the fledgling up and make sure it is not injured; if it is not, we return it to the nest. Usually the second attempt at flight goes well.

LANDING

When returning to a perch, Peregrines may slow their speed by flying below the landing spot and arcing up towards it. Wings are spread wide and back to further reduce velocity. Legs are extended for landing.

CHAPTER NINE

Prey

ARTIST NOTE: *The subject of prey offered me an opportunity to render bird specimens lying on a tray and therefore a strongly foreshortened subject. Now this is not an issue when drawing from life because when you locate and measure distances, the image will flatten and be easy transferred to a flat page. Hold up your pencil horizontally, extend your arm, and close one eye. When I let the pencil cross the bird's feet (second from the lower left), I could now see how close they were to the beak. When I held the pencil vertically and let it run down from the beak, I could see that the feet were below and just to the right of the beak. That is finding location.*

As Peregrines are aerial hunters, they capture most of their prey while in flight. The stoop, in which falcons dive from above towards the prey, is the best-known mode. They will also engage in direct pursuits, chasing in an attempt to overtake and grab the prey. High-speed photography has shown that Peregrines grab their prey with the foot open and immediately close their fist after capture. There is of course variation in both the method of pursuit and type of capture depending upon the situation and prey species being pursued.

Scientists can study what Peregrines are feeding on by looking at prey remains that build up at nest sites. This can give us information on avian species distribution and migration. Also, feathers have a chemical fingerprint in their isotopes that allows scientists to figure out where the bird was when it grew. This may lead to new discoveries on topics such as migration patterns or wintering sites.

FIELD MUSEUM
NATURAL HISTORY

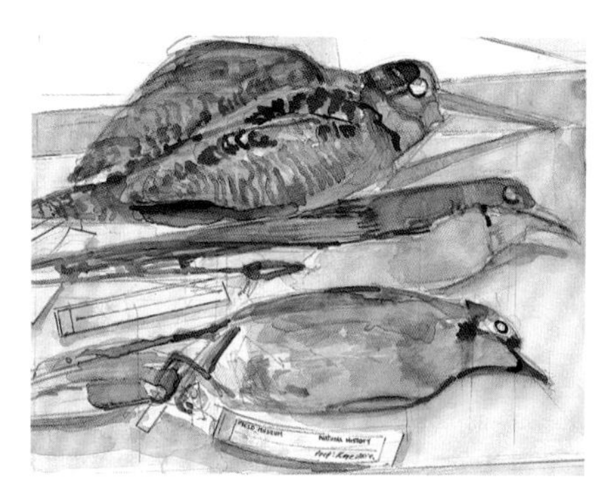

Prey
Top: American Woodcock (*Scolopax minor*)
Middle: Yellow-billed Cuckoo (*Coccyzus americanus*)
Bottom: Blue Jay (*Cyanocitta cristata*)

One common feature of the most frequent Peregrine prey species is that they are straight flyers. That, coupled with a significant biomass, makes these three species cost-productive to hunt, meaning from the perspective of a Peregrine, they make a great meal.

Blue Jay
Cyanocitta cristata
Order: Passeriformes
Family: Corvidae

Blue jays frequently mimic the call of hawks. These calls may announce to other jays that a hawk is in the area. Another theory is that the calls are used for kleptoparasitism. In other words, jays are trying to fool other birds into believing a hawk is in the area, scaring them so that the jay can steal their food. Blue Jays caught by Peregrines in Chicago are usually migrants.

Prey Feathers
1. Cedar Waxwing (*Bombycilla cedrorum*)
2. Chimney Swift (*Chaetura pelagica*)
3. Yellow-bellied Sapsucker (*Sphyrapicus varius*)
4. Yellow-billed Cuckoo (*Coccyzus americanus*)
5. Eastern Bluebird (*Sialia sialis*)

Chimney Swift
Chaetura pelagica
Order: Apodiformes
Family: Apodidae

During migration, thousands of swifts roost together in chimneys, funneling into them at dusk in spectacular tornado-like flocks.

Scientists have documented over 450 avian species taken by North American Peregrines as prey. By correlating prey remains to known migration dates, we can show that Peregrines take advantage of bird migrants passing through the area. During Chicago winters, we often see Peregrines move their territory towards open water, attracted by wintering waterfowl. Their diet will also shift in a higher proportion to species such as pigeons.

CHAPTER TEN

Banding

SCIENTIST NOTE: *This book could be filled with stories about banding, and I think among everyone's favorites would be those that involved children. They're our future scientists. Reaching them at a young age and seeing them excited about nature is a special thing. Kids are unashamedly honest, imaginative, and possess endless enthusiasm for something they love.*

We often thank the people we work with (building workers, monitors, students, etc.) by asking them to come up with names for the Peregrine chicks. Two of my favorite names have come from children—Banana Peel and Marshmallow.

Banding young Peregrines provides an opportunity to document the birds' movements and longevity.

Nestlings are briefly removed from accessible nest sites in order to place on them aluminum bands that wrap their legs like a bracelet. Unique to the individual, if the falcons are observed later, or found injured or dead, age and location of the bird can be recorded. This information can be used to understand dispersal patterns as well as age-related data such as breeding age.

While having the young Peregrines in hand during banding, scientists check for overall health and condition. A small amount of blood is collected from the chicks. They also save any unhatched eggs, egg shell fragments, and prey remains for later examination. When possible, this is also an excellent time to invite the public and/or press to observe the banding process and educate them about the species.

Protecting
Banding

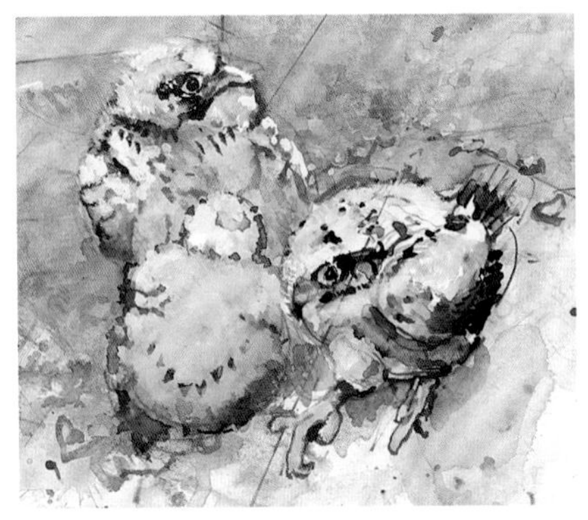

BANDING AGE

Banding is done when nestlings are around twenty-one to twenty-four days old. At this stage the chicks have reached full physical growth. Identification bands can be placed on the birds' legs without worry that the bird is going to outgrow the band. If there is any question on sex, a larger female band would be used. Also, since the chicks do not have their flight feathers, scientists do not have to worry about an accidental premature fledging from the nest.

CHICK GRAB

Adult peregrine falcons, particularly the females, are quite aggressive in defense of their young. Individual birds vary in their level of aggressiveness. They commonly fly very close to the bander but the more determined individuals may even strike them with their feet!

Scientists must deal with the defensive adults while retrieving the chicks for banding. Inverted brooms are used as a means of blocking the Peregrine's strike. Never used to swing at or hit the falcons, the brooms are held aloft. Peregrines like to strike at the highest point on the predator. Brooms are safe for the falcons because if the birds do strike, their feet easily pass through the soft bristles.

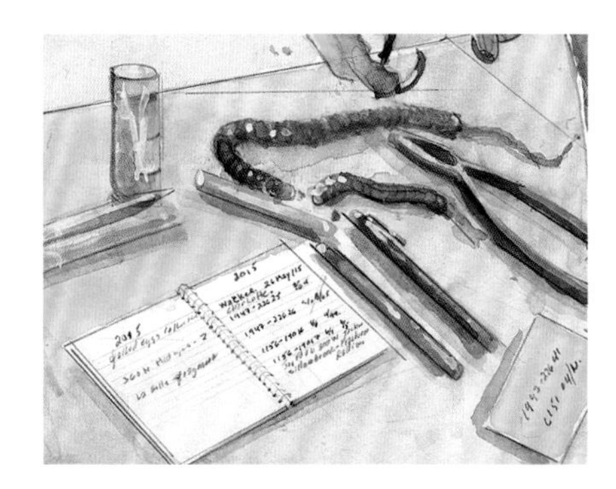

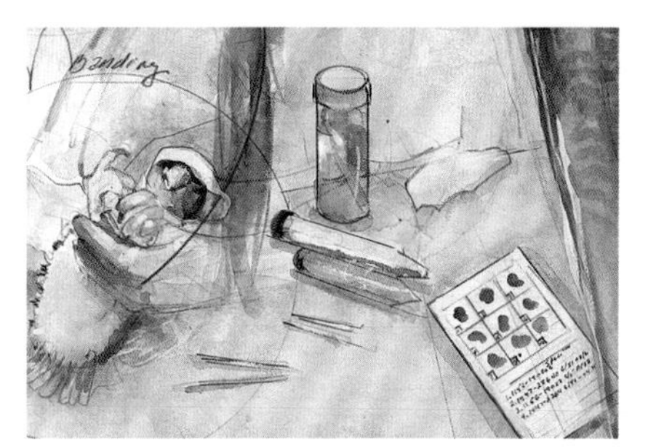

BANDS

As bird bands are unique to the individual, with return sightings scientists can study longevity and dispersal. Two bands are used for the Peregrines. The United States Fish and Wildlife Service (USFWS) band has a nine-digit number that wraps around the circumference of the band. This can be difficult to read unless the bird is in hand. Hence an auxiliary band that is more readable through a scope or camera is used.

Male Peregrines are roughly a third smaller than females and thus have smaller bands.

BLOOD DRAW

Scientists use banding as an opportunity to gather as much information about Peregrines as possible. By drawing a small sample of blood, scientists have studied genetics of the individual. They can also look for avian blood parasites or diseases.

CHAPTER ELEVEN

Research

SCIENTIST NOTE: *Dr. Isabel Caballero conducted her dissertation research at the University of Illinois at Chicago (UIC) on the genetics of Peregrines. Using molecular tools, among other things, she has provided insight into the genetic makeup, parentage, dispersal patterns, and breeding behavior of Illinois' Peregrines.*

While her work went beyond the scope of the Chicago Peregrines and Illinois, it was the details on "our" falcons that I frequently heard her get asked about. With unbanded adults, assessing parentage of offspring cannot be done through field observation alone. Looking at genetic markers, Isabel was able to tell us whether an unbanded adult was the biological parent to the chicks he or she was raising. Also, she could determine whether an unbanded adult was the same individual breeding at a particular site, year after year.

Peregrine recovery efforts provide unprecedented opportunities to study the species in depth. Illinois Peregrines have been part of various scientific studies for the past thirty years. Banding individuals allows scientists to look at longevity and dispersal patterns. Blood draws have provided data for examinations of Peregrine genetics. Chicks have been examined for avian ectoparasites. Addled eggs are collected each year for ongoing studies of shell thickness and contaminant levels. Remains of Peregrine prey are examined to survey the diversity of the species hunted as well as how migration can play a role in the Peregrines' diet. These are just some of the research efforts being conducted on Illinois' Peregrines.

In addition to scientists, the general public also plays a role in enhancing our knowledge about Peregrines. Monitors are more than just observers. Their observations recording individual identities, movements, and behaviors contribute to the breadth of knowledge on the species.

OÖLOGICAL COLLECTION
Doctor M.T. CLECKLEY
Date
Mo Day Yr
Set Mark
Eggs in set
Name
Town
Identification
Birds
Nest Situated
Diameter Inside 12 inches
Outside
Collected by
No 126
Research on Peregrines

Extra-pair copulations

Carnidae flies

Ectoparasite collecting
Photo: S. Ware, 22 May 2007, Pilsen

From genetics, we can determine whether the adult Peregrines raising a brood of chicks are the biological parents of their offspring. With higher population numbers, there are a greater number of Peregrines that do not currently hold a territory but rather "float" around the population waiting for the opportunity to breed. The question arose on whether one of these lone birds could have snuck in to breed with the resident female. If so, the young would be a result of extra-pair copulations. Illinois has not seen evidence of this to date.

The only known occurrence of an unrelated adult raising young was at the Wacker site. In 1998, the resident adult female was injured early in the year. A new female paired with the resident male and laid four eggs. Another female, that was later identified incubating the eggs and raising the two young, was not biologically related to the offspring.

Besides the usual banding and collecting of blood for genetics, we took the opportunity to help with an avian lice study. Jason Weckstein, an associate researcher at the Field Museum (2010–14) was studying the evolution of bird parasites. We attempted to collect any "bugs" we found on the chicks. While no lice were found, we did collect adult flies in the family Carnidae (genus *Carnus*), which are blood-sucking parasites of the nestlings that do not adversely affect the Peregrines.

CHAPTER TWELVE

Education

SCIENTIST NOTE: *Over the past thirty years I've given a lot of educational programs telling the Peregrines' story. One occasion stands out above all the rest. I was giving a program at Ryerson Conservation Area as one of a series of environmental talks throughout a daylong event. Their keynote speaker for the evening was Dr. Lester Fisher, former director of the Lincoln Park Zoo (LPZ).*

When I was a kid growing up in the Chicago suburbs during the 1960s–70s, everyone watched The Ray Rayner Show. *Periodically, Ray would have a segment called "Ark in the Park," where he would visit LPZ and Dr. Fisher would talk about a particular animal. I grew up loving that show, especially the trips to the zoo. I was fortunate to have met Dr. Fisher briefly as an adult when an injured Peregrine was brought to the zoo. I didn't speak to him but watched awestruck while he talked to the vet.*

At the start of my Ryerson program, I related that story, laughed at my "shy fan reaction to seeing Dr. Fisher," and encouraged everyone to make sure they went to his talk that evening. No sooner had I said that and begun my actual program but who walked into the cabin to sit down and hear me? Dr. Fisher himself! Unbelievable that an idol of mine from childhood would come to listen to me speak! It was astounding. To top it off, Dr. Fisher asked me to have lunch with him and I spent an enjoyable hour talking birds. Though I have to admit my first question was "How's Ray Rayner doing?"

Education is a vital component of the Chicago Peregrine Program. We reach the public using a variety of methods including the Internet and social media (website, Facebook, and various other sites like Flickr), public programming, and just daily interactions with residents, building personnel, and anyone that has to cope with the presence of Peregrines.

Social media have proven an excellent means to engage the public. The Illinois Peregrine Facebook page is followed by people in over fifty countries. The website has made information on Illinois Peregrines more readily available to others and provided an avenue for the public to interact with program personnel. This online data has proved to be an excellent resource for teachers as well.

Red-headed Woodpecker
Melanerpes erythrocephalus
Order: Piciformes
Family: Picidae

The Red-headed Woodpecker is one of four North American Woodpeckers that stores its food but the only one known to cover it with bark. They are also unusual in that they are able to catch insects while in flight.

Red-headed Woodpeckers have been documented as one of the more than a hundred prey species of Chicago Peregrines. Just as we can't use Peregrines to control unwanted birds such as nuisance pigeons, we can't stop them from feeding Red-headed Woodpeckers that we are looking to preserve.

The Red-headed Woodpecker is a species in the midst of a deep population decline. From 1966 to 2010, it is estimated that their population loss was 2.7 percent per year. The main problem is the disappearance of the mature beech and oak forests the woodpeckers depend on. The species was on the 2014 State of the Birds Watch List, which lists birds most in danger of extinction without significant conservation action.

Because Peregrines are a high-profile species, we can use discussions of the falcons to springboard the conversation towards other conservation issues such as habitat preservation of forests, wetlands, and prairies. In doing so, perhaps we can educate the public about other species of concern such as Red-headed Woodpeckers and what we can do to help them.

One role of education is expanding the knowledge base of an individual or group. In the process of studying Peregrines, we have at times been able to learn about other species as well.

Peregrines are predominantly aerial hunters, catching their prey while in flight. While the majority of prey is birds, they will occasionally feed on bats. One morning at the Illinois Beach State Park hack site, one of the immature Peregrines flew out over the Lake chasing something that flew like a butterfly. Soon a few of the other Peregrines joined the chase, returning to the beach to eat. It turned out they were feeding on migrating bats. Examining the prey remains, we were able to document three species of bats migrating together during the day.

Bat Species documented as prey by Illinois' Peregrines
Silver-haired Bat (*Lasionycteris noctivagans*)
Big Brown Bat (*Eptesicus fuscus*)
Red Bat (*Lasiurus borealis*)
Hoary Bat (*Lasiurus cinereus*)

CHAPTER THIRTEEN

Peregrine Dispersal

SCIENTIST NOTE: *The most frequently asked question we receive from anyone who follows the Peregrines is "Do you know where XXX is?" In truth, we hear about a very small number of birds after they have dispersed out of the natal territory. Raptor mortality is high with an estimated 60 percent not living through their first year. An unfortunate injury or death (if found) does provide an opportunity to have the bird in hand and verify band information. While that is helpful, happily a large majority of identifications occur with adult breeders at nests. Peregrines remain within a small area for months raising their brood. Unlike at cliff sites, in an urban location the birds and their observers can be quite close, increasing the odds of reading bands.*

It is extremely satisfying when we can occasionally answer that question of where XXX goes. For example, we've received reports on where a number of the young born at the Waukegan, Illinois, site have chosen to breed. Mapping out the locations, we can show how the young have followed the Lake Michigan shoreline, setting up various nest sites in urban areas.

One benefit of having Peregrines banded is that scientists can study where the birds travel to and from. When Peregrines disperse (leave a specific area), the direction may be towards a particular place or it may involve wandering. Young birds are looking for a place to find their first mates.

The Chicago Peregrine Program often receives valuable information about Peregrine whereabouts from birdwatchers, building personnel, and other concerned citizens. Once individual birds are located, scientists can study their movements. If you happen to see what you believe to be a Peregrine, try taking a picture even with a cell phone and send it our way. Don't forget to try and photograph the legs in case they have bands!

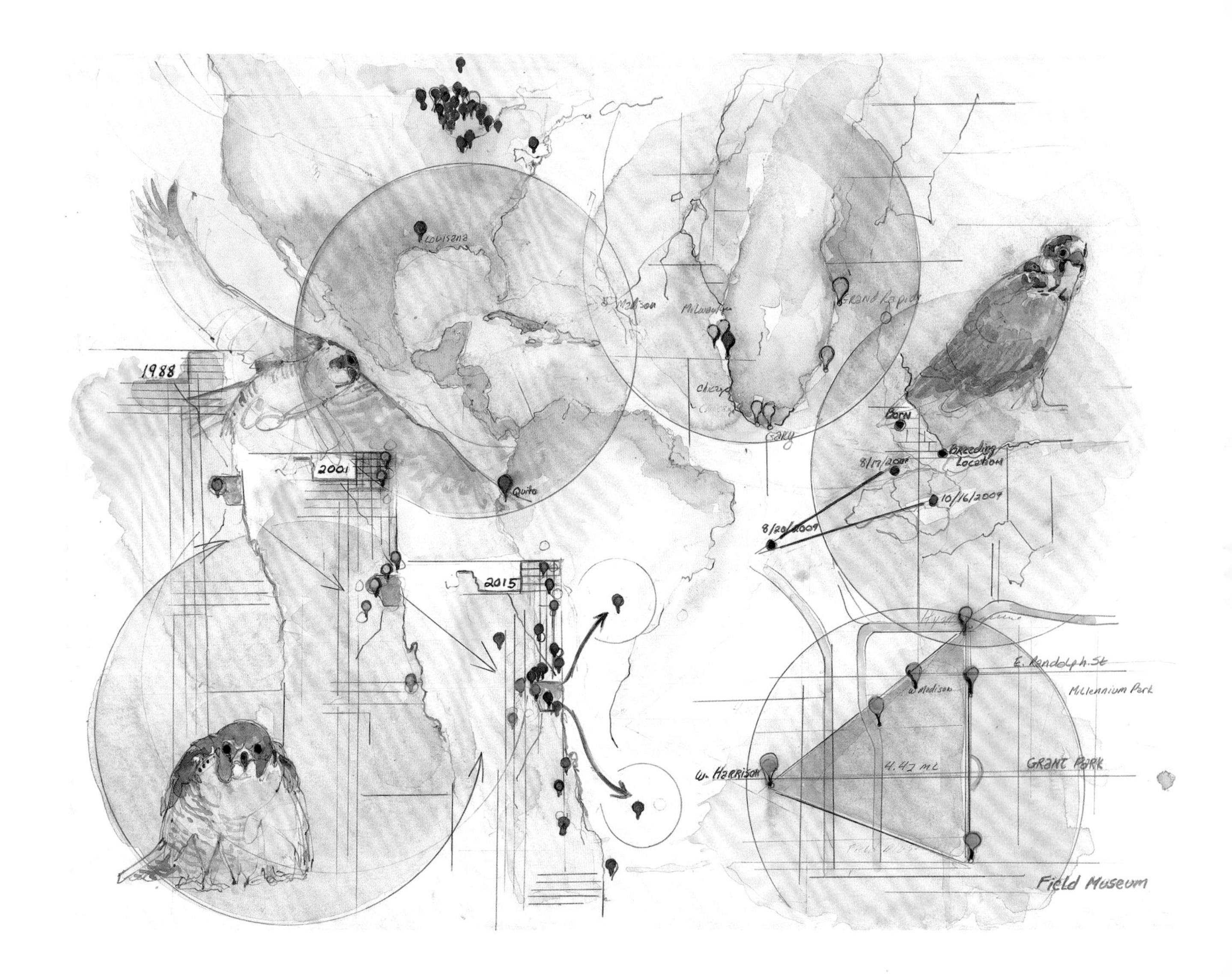

Louisana
Quito
Madison
Milwaukee
Grand Rapids
Gary
Born
Breeding Location
8/17/2008
10/16/2009
8/20/2009
1988
2001
2015
E. Randolph St
Millennium Park
W. Madison
W. Harrison
4.42 mi
Grant Park
Field Museum

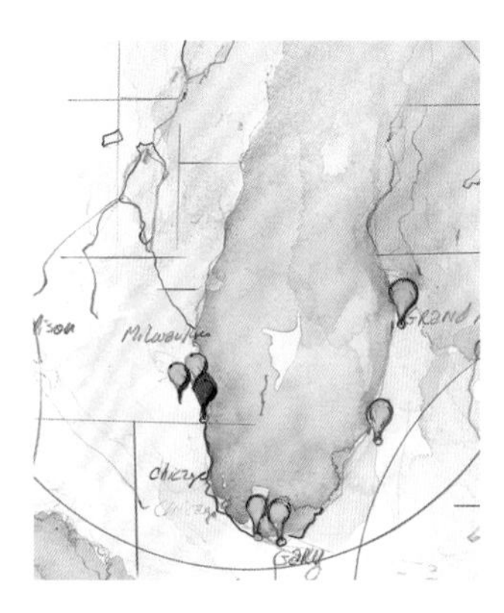

Waukegan Young breeding locations

Etienne: the male who traveled the farthest to reach Illinois
Photo: S. Ware, 4 May 2007, Wacker site

Etienne
Breeding years: 2006–7
Band identification: blk/blk 7/6, 816-84747
Natal site: Etobicoke, Ontario, Canada (2002)
Distance to Wacker site: 430 geodesic miles

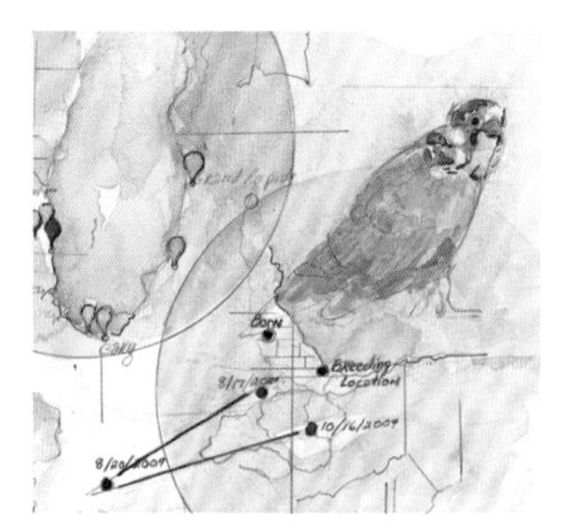

Dispersal pattern: natal to breeding

Recent studies on Midwestern banded Peregrines show mean natal dispersal (natal site to breeding location) for male Peregrine Falcons is roughly half that of females. The male that traveled the farthest to reach Illinois was Etienne.

In regards to the direction of dispersal, although we have seen birds move in all directions, the greatest tendency was a southeast movement. This tendency is reflected in the repetitive natal location of some of Chicago's breeding adult Peregrines. Over the years, five adults having fledged from the City Center building in Minneapolis relocated to Chicago to breed. Those individuals were Harriet (fledged in 1985, bred at Wacker 1987–97), Hubert (born 1985, bred at Wacker 1992–2004), Hercules (born 1997, bred at MCC 2006–9), Marriot (born 2001, bred at Calumet 2011), and Marquette (born 2007, bred at McKinley 2010–11).

On their own, young Peregrines will explore the Midwest before migrating out of the area for the winter. Occasionally some will wander out of the region. One Illinois offspring traveled as far east as New York City, and another as far south as Ecuador. On rare occasions, some immature Peregrines opt to not move at all and instead stay with their parents throughout the winter. Come spring, the adults will chase their last year's offspring away as they get ready to breed again.

Unlike most immature Peregrines, Illinois' adults stay year round. Documenting these movements (or lack thereof) answers an interesting question scientists had regarding genetically versus geographically driven behavior in Peregrines. Historically, the subspecies in Illinois was nonmigratory. To have the stock to release, scientists crossbred a number of subspecies, both migratory and nonmigratory. The question was whether the reintroduced falcons would migrate. The answer is that most current Illinois adult Peregrines do not.

CHAPTER FOURTEEN

Cultural Nest Locations

ARTIST NOTE: *A little bit of perspective drawing goes a long way. When rendering a building like the Evanston Public Library, I begin by determining the horizon line. This line is horizontal and not angled. It is where the eye rests if looking straight ahead. Next I draw the strongest angle of the building and extend this until it meets the horizon line. Where they meet I have the vanishing point and all the other parallel lines will go to this point.*

Libraries and universities, as advanced institutions of learning, are often thought of being at the opposite end of a cultural spectrum from a prison. Peregrines make no such distinction between them when choosing where to roost or nest. While the human use of the inside of a structure has little meaning for Peregrines, activities on the outside can influence Peregrine behavior during the breeding season.

Nest location: Evanston Public Library

Year Open (current site used by Peregrines): 1908
Young fledged: 42

Squawker: adult male
Photo: S. Ware, 12 May 2009, Evanston Public Library

Squawker
Band identification: b/g 48/M, 2206-49420
Natal site and year: Pleasant Prairie, Wisconsin, 2003
Breeding location and years: Evanston Public Library site, 2006–15

Magnolia
Photo: S. Ware, 29 May 2007, Hyde Park, Chicago

Magnolia
Band identification: blk 22R, 1807-29444
Natal site and year: LaCrosse, Wisconsin, 1991
Breeding location and years: Chicago Hyde Park sites, 1994–2009

To date, Magnolia is Illinois' oldest breeder, as she was still laying eggs at age eighteen.

University of Chicago, Hyde Park

Year open: 1892
Young produced: 0

The University of Chicago and the Evanston Public Library are both important cultural institutions. One location successfully produces Peregrine young on an annual basis. The other has failed in each attempt. What is the difference?

One problem is that Peregrines at the university favor gutters. While a gutter may have all the features falcons are looking for, ultimately the nest will wash out after a heavy rain. The pair at the Evanston Library first used a flower box and later moved up to the top of a structural column. This site has an added protection of an overhanging roof that protects the nest site from weather. It has an advantage for the scientists, as the roof prevents the adults from being able to fly at and strike when the chicks are retrieved for banding. The greatest challenge at the Evanston site is during fledging. Because of close proximity to the street, a number of young become grounded.

Metropolitan Correctional Center site

Site Information

Young fledged: 38 (1998–2014)

Year opened: 1975

Architect: Harry Weese

Immature Peregrines near fledging age

Photo: S. Ware, 12 June 2006, Metropolitan Correctional Center

A unique set of challenges accompany monitoring Peregrines at a federal prison facility. Security is understandably high. Banding would be a challenge. You cannot rappel from the top, since the roof is an exercise yard for prisoners. You cannot exit a window to rappel, since windows are extremely narrow to prevent escapes (which hasn't always worked.) The nest ledge is far enough away from the street to prohibit access with a cherry picker. So in this case, young are banded only if they become grounded. And how do you return the fledgling to the nest when the nest is inaccessible? Obviously you can't, so with the prison site we release any birds on the roof of an adjacent building.

CHAPTER FIFTEEN

Crib Peregrines

ARTIST NOTE: *The Wilson crib plate is a narrative about peregrines and their neighbors, as well as their unusual habitat. When I first began drawing at the Field Museum, I drew pots in the China exhibit for about ten years. A pot, like the crib structure can be divided in half longitudinally. Then you use horizontal lines to stabilize each curve. It should be the same height on either side. Measuring (taking amounts) and comparing will help you adjust the strength of each curved surface.*

Chicago Peregrines have nested on islands two miles offshore into Lake Michigan: water intake cribs, designed to provide drinking water for the city. The inactive Wilson Avenue crib and the active 68th Street and Harrison cribs have provided excellent places to nest. The building structures on the islands provide the pseudo-cliff ledges that Peregrines prefer.

A nesting pair of Peregrines will not allow another Peregrine in their breeding territory, let alone nest alongside of them. This does not always hold true with Peregrines towards other bird species. At the Wilson crib in close proximity to the Peregrines are nesting Double-crested Cormorants (*Phalacrocorax auritus*) and Herring Gulls (*Larus argentatus*). Another example of this tolerance occurred at the Hyde Park site in Chicago. Peregrines nested on a ledge outside a church steeple while numerous pigeons nested inside. The pigeons often flew directly over the incubating Peregrine's head to reach their nests.

Double-crested Cormorant

Phalacrocorax auritus

Order: Suliformes

Family: Phalacrocoracidae

Double-crested Cormorants are widespread throughout North America, and most frequently seen in freshwater. They form colonies of stick nests built high in trees on islands or in patches of flooded trees.

Herring Gulls

Photo: S. Ware, 12 June 2014, Wilson Avenue crib, Chicago

Herring Gull

Larus argentatus

Order: Charadiiformes

Family: Laridae

Adult Herring Gulls are gray and white with pink legs and a yellow bill with a red spot near the tip of the lower half. Their plumage varies greatly over the first four years of life so identification can take some skill.

Can you spot the Peregrine in this photo?

Wilson Avenue crib, Chicago

Photo: S. Ware, 12 June 2014

Site Information

Active Peregrine years: 2013–15

Adult ID, female: Chayton (b/r 80/H)

Adult ID, male: unbanded

Young produced: 10

Unlike Peregrines, Double-crested Cormorants and Herring Gulls are colonial breeders. Groups of nesting birds of either species may be found in diverse habitats such as fresh water lakes, islands, and coastal areas, for example. Active cribs are free of these species due to the human activity. On the other hand, birds at the inactive cribs have free rein of the island.

Peregrines have shared the Wilson crib with nesting Double-crested Cormorants and Herring Gulls since 2013. Work on the crib began in 1915 with eight miles of tunnels hand dug in the bedrock beneath Lake Michigan. The light at this site serves as a navigational aid for boaters.

The 68th Street crib consists of two small islands connected by a walkway. Peregrines nested on a small ledge under the walkway. This crib was the site of a tragic fire in 1909 in which an estimated number of over forty men constructing the crib lost their lives.

Peregrines have been attempting breeding at the three of the four water cribs in Chicago. In July 2015, one of the young from the 68th Street crib was found dead near Gary, Indiana. Though sad news, it does confirm the Peregrines are able to safely fledge off the crib islands, two miles offshore.

68th Street crib, Chicago
Photo: M. Hennen, 23 May 2015

Site Information
Year active: 2015
Adult ID, female: unbanded
Adult ID, male: unknown
Young produced: 2

CHAPTER SIXTEEN

Landmark Buildings

ARTIST NOTE: *Watercolor affords the artist an added tool of transparency. If you stay away from the Cadmiums, most colors can be reduced to at least a semitransparent state. In that state you can redraw over a finished dry area, and highlight a specific element. It is safe to leave the cadmiums until the end of the process, when you will need to makes areas pop.*

Once a building is designated a landmark, it is subject to the Chicago Landmarks Ordinance, which requires that any alterations beyond routine maintenance be reviewed by the Landmarks Commission. If listed on the National Registry of Historic Places, federal tax support for preservation is available.

Though Peregrines and their eggs and/or young are protected, their presence on a landmark building does not add any extra level of protection for the structure.

SEARS ROEBUCK

Nitz
Photo: S. Ware, 5 June 2015, University of Illinois at Chicago

Nitz
Band identification: b/g 2/*Y, 1807-77765
Natal site and year: Milwaukee, Wisconsin, 2001
Breeding locations and years: Old Sears Tower, Chicago, 2006; University of Illinois at Chicago, 2014–15

Old Sears Tower site, Lawndale, Chicago

Site Information: Old Sears Tower
Year of Peregrine activity: 2006
Year opened: 1906
Architect: Nimmons & Fellows

From 2004 to 2007, Peregrines called the Old Sears Tower in Chicago's Lawndale neighborhood home. Breeding was confirmed for only a single season (2006), when an infertile egg was collected. The female, Nitz, fledged from a nest in Milwaukee in 2001 and resided at Lawndale from 2004 to 2007. The site underwent renovation, and eventually Nitz abandoned the territory. In 2014, at the age of thirteen, she reappeared at the University of Illinois at Chicago and replaced Rosie, the resident female. It is unknown where Nitz was during the interim years.

When the Powhatan building was actively being used by Peregrines, it hosted one of only three breeding pairs in Illinois. The falcons used a nest box placed on the upper roof, which had been installed after Peregrines were observed hanging out on the fire escapes with no suitable ledges for nesting. Difficulty arose when the roof just below the nest box had to be closed off to apartment residents to protect them from the birds, who wanted to defend their nest. In 1995, while the majority of the residents were willing to sacrifice their roof for the forty-five days it would take the young to fledge, a few were adamantly against hosting the falcons. The decision was made to remove the box after the nesting season to ensure the future safety of the Peregrines. Fortunately, the birds relocated at a new site 1.5 miles away.

In downtown Elgin, Peregrines frequented a historic office building from 2011 to 2013. The Elgin Tower building was constructed in 1929 to house the Home National Bank and Home National Savings and Trust and is one of only two art deco buildings in the city. Peregrines failed in their only attempt at breeding at this site. In 2011, three eggs were laid on an outer ledge and failed to hatch. An arson fire on May 4, 2014, destroyed parts of the interior of the building and perhaps was the reason the Peregrines abandoned use of the building even as a roosting site.

Peregrines are often seen along the Elgin bridges over the Fox River.

Powhatan art deco façade

Site information

Young fledged: 3 (1994–95)
Year opened: 1929
Architects: Robert De Golyer and Charles L. Morgan

Elgin Peregrine
Photo: S. Ware, 7 June 2011, Elgin

Elgin female

Band identification: b/r 95/H, 1687-30406
Natal site and year: Omaha, Nebraska, 2009

CHAPTER SEVENTEEN

Uptown

SCIENTIST NOTE: *The first year we banded at Uptown is the one and only time we had the Illinois Department of Natural Resources conservation police called on us for our work with Peregrines. And it was entirely my fault. Not for harming the falcons but because of my fascination with the theater.*

While retrieving the chicks for banding, a few of the neighborhood residents who had been watching the Peregrines saw us remove the chicks from the nest. It was great to hear their concern for the birds as they yelled for us to put the birds back. My plan was talk to talk to them right after we finished banding. Normally it would have only taken twenty minutes max. This time is it took over ninety minutes before we departed the theater and the police were waiting at the door!

Why did it take so long? Once we finished banding and returned the chicks to the nest, the theater caretaker offered to give us a tour. It was an opportunity I couldn't pass. He spoke of a rich history, from the ornate furnishings and elaborate design to the parade of stars that crossed the stage. By far my favorite part was the nursery, where everything was built to scale for children, including the carousel they could ride. Remnants of the circus theme still adorned the walls. I didn't need the partially restored section of the lobby to picture how the theater looked like in its heyday. In my mind I saw it all.

We did eventually make it out of the theater to show our permits and explain to the conservation police (and concerned citizens) what we had been doing with the Peregrines. Their only request was next time to be invited along!

Chicago's Uptown Peregrines resided on the backside of an old movie palace, the Uptown Theater. Opened in 1925, the Uptown is a grand theater that occupies 46,000 square feet and has 4,381 seats. Unfortunately, the theater closed in 1981 and later suffered severe damage when it remained unheated through a brutally cold Chicago winter.

For the entire time that the site was active (2001–13), only one female (Zoom) occupied the territory. This is unusual; in Illinois' recent Peregrine nesting history most sites see a changeover of adults for each sex. Zoom went through several males and spent the last year on her own. During the nonbreeding season, the Uptown Peregrines could often be found a mile to the east at Montrose Harbor. Since Zoom disappeared, the eyrie has been empty, though an occasional Peregrine is seen at Montrose. Surprisingly, as of 2015, no other Peregrines have used the theater's nest box.

Chicago's Uptown neighborhood was home to a pair of Peregrines for over a decade. Their eyrie was located on a fire escape of an old movie palace, the Uptown Theater. The situation at Uptown was such that the building framework of the door created a small ledge the falcons could use. In a case such as at the Pilsen fire escape site, the Peregrines used an old pigeon nest whose materials had created a solid ledge within the grating. When the nest location is susceptible to failure, one can install a nest box, which increases the chance of a successful nest. This was done at both the Pilsen and Uptown sites.

Uptown Theater, Chicago
Architects: C. W. and Geo. L. Rapp

Site Information: 2001–13
Total eggs laid: 43
Eggs hatched: 34
Young fledged: 32

Uptown pair
Photo: S. Ware, 29 May 2009

Zoom (front)
Band identification: b/r *4/H, 1807-61930
Natal site and year: St. Paul, Minnesota, 1997
Breeding years at Uptown: 2001–13
Breeding years at Evanston First United Methodist Church: 1999

G/G (behind)
Band identification: b/g G/G, 2206-35789
Natal site and year: Michigan City, Indiana, 2001
Breeding years at Uptown: 2003–11

Inbreeding at Uptown Theater

Stan
Photo: S. Ware, 10 June 2013

Stan
Band identification: b/g 98/K, 2206-62900
Natal site and year: Uptown Theater, Chicago, 2005
Breeding locations and years: Belmont/Addison, Chicago (2007), Uptown (2011–12), Broadway (2013–15)

Inbreeding is not easily documented in wild animal populations due to the difficulty in identifying individuals and obtaining genetic information. Because the Midwest Peregrines have been banded and blood has been drawn from each individual, scientists can document such occurrences. Illinois has recorded two pairs of related Peregrines breeding with each other. The first was in Hyde Park (2007–9) and the second at the Uptown Theater (2011–12). At Uptown, the adult female (Zoom, b/r *4/H) nested with her son from 2005, Stan (b/g 98/K). They successfully raised young each year before Stan left for another site.

It is estimated that roughly 4 percent of the Midwest Peregrine population is from related adults (full siblings, half siblings, and parent with offspring). With the high percentage of marked individuals (through banding) and blood samples taken for genetics, a more exact record of occurrence of inbreeding in Peregrines may be found.

CHAPTER EIGHTEEN

Industrial Sites

SCIENTIST NOTE: *Anyone who follows the Chicago Peregrine Program (like us on Facebook) knows Stephanie Ware and Josh Engel, vital members of the team. Stephanie's photos of Peregrines are extraordinary. No contest. But one of my favorite photos happens to be a picture I took in 2015 while banding at our Pilsen warehouse site. Stephanie was "blocker" for Josh as he went out to grab the chicks. When the fire escape door opened, a wind vortex was created, causing all the prey feathers in the nest box to swirl around Josh. The look on both Josh's and the falcon's faces—priceless!*

Traditionally industrial buildings had to be placed along a waterway or railroad as a means of moving their product. Peregrines looking for their cliff nest may follow waterways (river or lake edge) in search of a perfect ledge. It makes sense that the falcons would discover an artificial cliff on a warehouse or power plant.

Some states used boxes placed on power plants as an incentive to entice Peregrines to nest. Unless requested by a specific site for a nest box, Illinois used a more reactive approach, preferring to let the Peregrines come first. The Pilsen site is an excellent example. Peregrines discovered the warehouse on their own, utilizing a grating on a fire escape. The nest box was installed later to give the Peregrines a safer spot to use.

In a few instances, Illinois did install boxes prior to the presence of falcons. It was a 50/50 shot as to whether they would ever be used. Sometimes Peregrines never came and the box eventually was taken down. Some places, like Waukegan, had their box up for years before being occupied by Peregrines. And then others, such as Savanna, had a pair of Peregrines move in immediately.

MIDWEST
GENERATION
THE Field
MUSEUM
Midwest
GENERATION

Savanna adults
Photo: S. Ware
3 May 2012, Savanna Site

Site Information
Young fledged: 14 (2010–2014)
Male and female are both unbanded

Savanna site: Consolidated Grain & Barge

In 2010, the U.S. Fish and Wildlife Service erected a nest box on a grain elevator in Savanna, Illinois. Immediately, a pair of unbanded adult Peregrines began using the box. While the majority of Illinois' adult Peregrines remain year round, the Savanna pair migrated out of the area in winter. Because these individuals are unbanded, we cannot confirm whether the movement is away from their nest site but still localized, or whether the move is of a significant distance.

Fran in flight
Photo: S. Ware, Waukegan, 3 June 2013

Fran
Band identification: b/g 5/*X
Natal site and year: Hoan Bridge, Milwaukee, Wisconsin, 1999
Breeding location and years: Waukegan site, 2001–13

NRG plant, Waukegan

Waukegan has proven to be one of the more productive Peregrine sites in Illinois. Forty-nine young have fledged from there. One young male was sighted along the Ecuadoran coast, 3,158 miles south of his natal site, the winter he fledged.

A number of Fran's offspring have headed to Indiana to set up their breeding sites. At one particular location (Whiting, Indiana), genetically full siblings born from different Waukegan broods have mated for at least a single season. Nancy (b/g 5/*4, 2002 offspring) bred with Hughes (b/g E/09, 2006 offspring) in 2012, fledging two young. Nancy was injured in December 2012 and deemed un-releasable. The following January, one of Nancy and Hughes' offspring (b/r 78/D) spent her first winter 2,200 miles south at a church in Costa Rica.

Pilsen site, Chicago

Site Information: 2002–2015
Total eggs laid: 44
Eggs hatched: 24
Young fledged: 21

Huff in defense
Photo: S. Ware, 10 June 2014, Pilsen site, Chicago

Huff
Band identification: b/r 24/N, 1126-14307
Natal site and year: Broadway site, Chicago, 2011
Breeding years at Pilsen: 2014–15

Chick Grab
Photo: M. Hennen
4 June 2015, Pilsen Site

In 2002, the Chicago Peregrine Program was contacted about a pair of Peregrines with chicks nesting on a warehouse in the Pilsen neighborhood. The falcons were using an old pigeon nest at the top of a fire escape. One characteristic of pigeon nests is that the material brought in to form the nest often includes twine. On June 19, one female chick was banded. The second chick had a leg injury (twine constricting the leg) and was taken to the hospital at the Lincoln Park Zoo. Unfortunately the injured chick had to be euthanized due to the infection in the leg bone resulting from the injury. Care was taken to make sure the nest site was free of any material that could potentially harm the remaining birds.

During the off season, a nest box was placed in the spot where the original pigeon nest sat. Peregrines have been using the nest box annually.

CHAPTER NINETEEN

Three of Chicago's Eyries

SCIENTIST NOTE: *Individual personality of a Peregrine contributes greatly towards its level of assertiveness as a defensive adult. Three main factors influence our ability to deal with defensive adult Peregrines: building design, number and position of balconies, and individual balcony design. When the architecture of the building can be used as a blocker from the adults, this is an advantage. For instance, the column in the center of the nest ledge at our UIC site prohibits the adults being able to strike the scientists when grabbing the chicks for banding.*

The key time to minimize human activity on a natal building is after the eggs hatch until fledging. This forty- to forty-five-day period is when the birds are the most defensive. We take care to talk to residents and engineers about what they need to be careful about. If the building, such as the Lakeview site in 2012, only has a single balcony, then any danger outside of the natal ledge is limited to the roof or fire escape (if present). This type of site generally has limited outside human activity and thus is easier to handle. Buildings with multiple balconies are more problematic because more aggressive birds will defend all balconies surrounding the natal site.

Peregrines have used buildings throughout Chicago, from north of Hollywood Avenue to the south side bordering Indiana. These sites have grown in number and fluctuated in location over the past thirty years.

Activity at each location must be verified annually. Scientists reconfirm the identity of the adult Peregrines when checking for nesting. Sometimes, even though everything remains the same with the falcons, personnel within the building have changed. This is why education is a vital aspect of any Peregrine work, as each year we must teach people how to live with the birds.

In total, nearly fifty buildings in the city alone have been used as nest sites. We would love to highlight each and every one but have instead chosen to present a few with interesting stories.

University Hall, University of Illinois at Chicago

Building Site Information
Year opened: 1965
Architect: Walter Netsch

Nesting Site Information: 1998–2015
Total eggs laid: 63
Eggs hatched: 43
Young fledged: 38

Adult male, UIC: Mouse
Photo: S. Ware, 5 June 2015

Mouse
Band identification: b/g W/72, 816-38648
Natal site and year: Broadway, Chicago, 2008
Breeding years at UIC: 2013–15

Broadway site

Site Information
Breeding years: 1994–2015
Total eggs laid: 92
Eggs hatched: 71
Young fledged: 62

University Hall is where it all began for the Chicago Peregrine Program. The roof of this building is where Illinois' first hacked-out Peregrines were released in 1986. It was one of those first-released birds (a male named Jingles) that took up residence slightly to the east at the Wacker site in 1987 to become the state's first post-decline breeder. Because of the proximity of UIC to the new Wacker nest site, the hack site location was moved to Ft. Sheridan. University Hall hosted no Peregrines until 1996, when a single bird began to frequent buildings on the UIC campus. This individual bird paired up in later spring of 1998, and the two Peregrines bred on a ledge on University Hall the following year.

Eleanor
Photo: T. Gnoske, 1994, Broadway site, Chicago

Eleanor
Band identification: b/r 2/8, 1807-34872
Natal site and year: Sheboygan, Wisconsin, 1993
Breeding years at Broadway: 1994–2002

Demolition 2014
Photo: M. Hennen, 2014, Dorchester site, Chicago

Site Information:
St. Laurence Church on Dorchester
Young fledged: 3 (2014)

St. Laurence Church, Dorchester Street, Chicago

Year opened: 1911
Architect: Joseph Molitor

When the Broadway site was first established as a Peregrine territory (1989), it ran from north of Hollywood Ave south to Irving Park Road, a length of 2.5 miles. During the early years, the adults ranged freely between buildings at the north and south ends of the territory. In 1994, a four-year-old male and a one-year-old female began nesting at Broadway. Peregrines normally begin breeding at two years of age. With low population numbers at the time, younger Peregrines such as one-year-old Eleanor had opportunities to nest. With current population levels, birds this age now rarely have the chance to breed.

Eleanor was replaced by a female named Auntie Em (b/g 5/*P), who consistently laid five eggs for eleven consecutive years of the thirteen years she was breeding. Auntie Em has been Illinois' most productive breeder to date.

St. Laurence Church was closed in 1999 and the building sold in 2005. After being left unattended for several winters, burst water pipes took their toll on the structure. The building was placed on Landmarks Illinois watch list (2007–8) and most endangered list in 2010. Though demolition was underway in the spring on 2014, a pair of Peregrines took up residence and raised two young in the abandoned steeple.

Cliffs are formed by a river slowed etching away the landscape, forming acceptable Peregrine breeding locations. Changes are made over time as new ledges form or wear away. The same may be said of city ledges, as buildings are torn down and new ones erected. Peregrines living in either a natural or artificial habitat must adapt to these changes.

CHAPTER TWENTY

Living with Peregrines

SCIENTIST NOTE: *Interactions with the public have run the gamut from enthusiasm to disdain, from serious to inane. We treat all requests diplomatically, no matter how silly we might find it. For example, we've had more than one request to throw out a flower pot because a Peregrine sat in it. We had a request to come and move a roosting bird on a balcony railing because the resident wanted to go out and smoke. Not all requests are negative towards the birds. One request was to find a way to put something over a nesting Peregrine as it was getting wet in the rain.*

We recently surveyed some of the various Peregrine building personnel, asking "How has your life/work been affected by Peregrines?" While aggressive adults can unquestionably complicate daily building operations, the net effect was overwhelmingly positive. The owner at the Millennium Park site remarked, "Cleaning up after them is a pain in the ass and they are always late with the rent, however it's fun having them around. :-) Seriously, it is nice to be able to make a difference for the species."

Is living with Peregrines easy, or not?

One role of the Chicago Peregrine Program is to be a liaison between the falcons and the public. Our job is to explain about Peregrines and to help the public in dealing with the species. Some of the key personnel we work with are building managers and engineers. Their job is to maintain the building, which can be compromised when there is a Peregrine defending its young from intruders. During the breeding season, the program works with residents and building personnel on how best to deal with the situation.

Another role of the program is to be a link to the falcons for Peregrine fans. We can answer questions and provide details on what is happening in the Peregrine world. It is a two-way street, as many individuals make our lives easier by being our eyes and ears.

Top: Unbanded male: Wrigleyville (Belmont/Addison), Chicago
Photo: S. Ware, 16 May 2013

Bottom: Balcony nesting

An unbanded pair of Peregrines has been holding a territory in Wrigleyville, a neighborhood between Belmont Avenue and Addison Street. In 2013, they nested on a private residence balcony. An empty flowerpot provided a safe place to lay their eggs. They fledged three young. While the condo owners were very accommodating to the Peregrines throughout the nesting season, after the birds left they opted to prevent any future use of the balcony by falcons. While the adults and young are protected, once the site has been abandoned at the end of breeding, owners are free to block Peregrines from using the site again as long as the method used will not cause any harm to the birds. This has happened with a number of Peregrines that have selected private residences as their nest sites.

While the Belmont/Addison pair failed in their three nesting attempts the following year, in 2015 they utilized a flower box at a new site and successfully fledged four young. Happily, their current host is welcoming to the birds and will make the box available for future use.

With an increased Peregrine population in urban centers like Chicago, it is increasingly common to find a private balcony selected as a nest site. This creates a challenge for all. The Chicago Peregrine Program assists the property owner, building management, engineers, and residents in dealing with the falcons.

The program's prime role in these circumstances is to explain how the nesting will progress as well as how best to cope with living with the falcons. All attempts are made to minimize any disturbance for the people and the birds. For example, one method to ease clean-up after the birds is to tarp the balcony.

Peregrines on private residences

Balcony tarping
Photo: M. Hennen, 2014, Oak Street site, Chicago

CHAPTER TWENTY-ONE

Another Opinion

SCIENTIST NOTE: *One of the greatest misunderstandings regarding Peregrines is the thought that scientists control where the birds go. It is not uncommon to get a request to come remove "your" bird. Part of the problem is that it is often written that Peregrines were "placed" in cities. That statement implies scientists can manipulate a Peregrine's behavior. Not so. To put it simply, Peregrines are wild birds, not tame or domesticated. They do what they want.*

Not everyone likes Peregrines. That's fine, not everyone has to. It is certainly an understandable reaction in situations when one has to deal with a defensive adult. It is unfortunate, though, when arguments against the falcons are full of faulty science. In that case we need to educate the public and correct any misunderstandings.

Another interesting twist is that not everyone who likes Peregrines agreed with reintroducing Peregrines back into the wild. Some scientists felt that since the original subspecies (*Falco p. anatum*) could not be restored, we should not introduce what would a genetic novelty by mixing a number of races (*Falco p. tundrius, Falco p. anatum, Falco p. pealei, Falco p. cassini, Falco p. peregrinus, Falco p. calidus*, and *Falco p. brookei*). To others, the problem lay with having releases in cities because they felt young Peregrines would imprint on urban settings to a point where they would not use historic cliffs.

The Chicago Peregrine Program has always tried to do what is best for the falcons but still assist people affected by the falcons' presence. If we can find a way to resolve an individuals' concerns, yet ensure that in the long run the Peregrines can continue to breed, then we've done our work right.

Birds should continue as a matter of biotic right, regardless of the presence or absence of economic advantage to us [humans].

ALDO LEOPOLD, *A SAND COUNTY ALMANAC*

Some people blamed Peregrines for wiping out localized populations of birds such as Mourning Doves. While Peregrines will indeed take doves as prey, in the totality of what a single pair of Peregrines consumes they do not eat very many. When asking why you may not see as many birds like Mourning Doves in your yard as you had before, there are a number of things to consider. First, has the neighboring habitat changed? Perhaps there are fewer trees and/or fewer people feeding the birds. Second, one must consider what other predators are in the area. Cooper's Hawks are more common than they were twenty years ago and they also prey on Mourning Doves. Feral cats kill billions of birds annually. Third, has the season changed and the birds migrated out of the area?

If a person is concerned about a raptor feeding on the birds at their feeders, the best recommendation is to stop feeding the birds for a while. The birds will disperse to other

Mourning Dove
Zenaida macroura
Order: Columbiformes
Family: Columbidae

Mourning Doves are small-headed, plump birds, found across North America. Buffy to tan in color, they have black spots on their wings and a pointed tail. The Cornell lab estimates the Mourning Dove population at 350 million even though on average over 20 million are harvested by hunters annually.

Rock Pigeon
Columba livia
Order: Columbiformes
Family: Columbidae

Pigeons are considered the number-one bird pest. They are found in cities throughout the world. Homing pigeons are famous for being able to find their way home from long distances and at great speeds.

feeders and the raptor will move on. Once the raptor is gone, put the feeders back up and the songbirds should return.

Peregrines and pigeon fanciers have a rocky relationship at best. Back in World War II, Peregrines were killed to prevent them from feeding on carrier pigeons that were used by troops to pass messages. In the U.S. after the decline of the Peregrine, some states had problems with having their release birds shot during their reintroduction process by pigeon fanciers concerned for their flocks. In the past decade, there has been an increase in the number of shootings of Peregrines. In England, the Royal Society for the Protection of Birds has counted at least fifty-four incidences of Peregrines being illegally killed in the last six years. The number of shootings in the U.S. has also risen.

Now we are starting to see it happen close to home. In 2014, a $5,000 reward was offered for information about the shooting of a Peregrine near West Allis, Wisconsin. Wisconsin lost three Peregrines in 2014 to illegal shootings. One of the individuals killed had fledged from an Illinois nest. In May 2014, Fran, the Waukegan female, died after a territorial fight. Upon examination, Fran was shown to be full of avian heartworm, which probably made her weak enough to lose the dispute. But buried within the breast muscle mass was a pellet. Fran had lived for quite some time after being shot, as a cyst had formed around the pellet. In March 2015, a resident Peregrine on the north side of Chicago was shot and killed.

Shooting Peregrines is illegal; we realize that the actions taken against the falcons were done by a few extreme individuals. Please report to the U.S. Fish and Wildlife Service any abuse of wildlife that you may know about.

CHAPTER TWENTY-TWO

City Wildlife

ARTIST NOTE: *Sometimes I am blessed with a beautiful mount. The taxidermist, Tom Gnoske, has a thorough understanding of bird anatomy and behavior. A well-constructed mount creates a totally lifelike look, which makes the artist's job easier. For the flying Snowy Owls I painted that appear later in this chapter, I used a fantastic photo by Arlene Koziol. Art is often collaboration.*

Preserving wildlife in an urban environment requires a delicate balance of harmonizing the protection of natural areas while enhancing human communities. Chicago supports many native ecosystems, from prairies to savannas and dunes, to woodlands and wetlands. Not only do Peregrines use the artificial landscape, but they may be found in the city's natural habitats as well. It follows that the falcons interact with wildlife in both settings.

Within Chicago, the project Chicago Wildlife Watch initiated a photographic study to better understand what wild species use the urban landscape. Four times a year, motion-activated cameras are spread throughout the city. Citizens are encouraged to view captured images online and assist with identification of any wildlife pictured. With the information gained, scientists can better understand the biodiversity of the region and how it utilizes the area.

Highlighted in this chapter are just a few of the wildlife species that Peregrine Falcons may interact with in the city.

American Crow
Corvus brachyrhynchos
Order: Passeriformes
Family: Corvidae

Crows are well known for harassing birds of prey. They will often mob a raptor, sometimes to chase it out of the area, sometimes to steal food. Crows are also known for stealing eggs, often carrying them off in their beak.

Will crows steal Peregrine eggs? Perhaps, but it's not likely to happen often. While crows might sneak in to take an egg if the eggs are unattended, an adult female Peregrine will not leave her eggs, even when a predator is present. The male Peregrine will take over incubating for the female for short periods of time. Also, Peregrines are very adept at chasing species they perceive to be a threat out of their territory.

Illinois Peregrines may come across Snowy Owls when the owls arrive for winter. One such encounter took place in late January 2012 at Northerly Island in Chicago. The owl had chosen to roost in what was the South Loop Peregrine's hunting ground. Upon sighting the Peregrine, the Snowy Owl adopted a defensive stance with wings spread. As the Peregrine swooped in to attack the owl, the Peregrine somersaulted to extend its talons towards the owl. After a couple of passes, the Peregrine flew off and the owl moved closer to the lake. Not all such encounters end in a favorable manner for either species.

Snowy Owl
Bubo scandiacus
Order: Strigiformes
Family: Strigidae

John James Audubon once saw a Snowy Owl lying at the edge of an ice hole, where it waited for fish and caught them using its feet. This owl breeds in the Artic and visits Illinois only during the winter.

Great Horned Owl
Bubo virginianus
Order: Strigiformes
Family: Strigidae

The Great Horned Owl is one of the most common owls in North America. You can find it in all types of habitats such as deserts, wetlands, forests, and grasslands, as well as in cities and backyards.

Raccoon
Procyon lotor
Order: Carnivora
Family: Procyonidae

Raccoons are noted for their intelligence. Vehicular injury is one of the most common causes of death. Raccoons are opportunistic in finding their dinner and places to den. They also have adapted well to urban areas.

Great Horned Owls have been cited as one reason released (hacked) Peregrines did not returned to cliffs but instead moved into cities. The owls also were a problem at some hack sites, including the last site used in Illinois. In 1990, a Great Horned Owl killed and ate one of the release falcons. While the attack was not witnessed, the ID band worn by the Peregrine was found within an owl pellet (a regurgitated ball of indigestible material that owls cough out after eating.)

Great Horned Owls are not the only problem cliff-dwelling Peregrines face. While examining ledges, scientists found evidence of raccoon presence at many sites. Even if raccoons did not impact original Peregrine populations, they clearly have the potential to affect any Peregrines attempting to use cliff ledges today. As raccoons are found throughout urban environments, they may be able to affect our city falcons even in seemingly inaccessible sites. The Peregrines that nested in Savanna from 2010 to 2014 had a raccoon remotely photographed in their nest box in 2015.

CHAPTER TWENTY-THREE

Not a Peregrine?

SCIENTIST NOTE: *Unless there's an accompanying photo, I dread the email or phone call looking for an identification of a bird. Others are far better than I at interpreting clues. But you can understand the difficulty when someone says, "It's brown and has streaks." Even if the rest of the clues indicate a raptor, that description fits Peregrines, Cooper's Hawks, female kestrels, Red-tailed Hawks . . .—you get the idea. One modern convenience that I do love is the cell phone camera. Snap a picture, email photo to the scientist, identification underway. Easy peasy!*

My all-time favorite mistaken identity came from an individual who rode into Chicago on the L train each morning. He contacted the Chicago Peregrine Program, believing he was spotting one of our falcons hanging about a church. Each time the train passed the church, the bird was observed perched in a new location. He heard about Peregrines nesting outside churches (and we did have a few pairs doing just that) and thought we should know. We of course thanked him for the report and later went to check out the structure. To our surprise, no Peregrines were sighted but the church roof was loaded with plastic Great Horned Owls!

The extent of the ability of many people to identify birds does not venture far beyond well-known species such as pigeons, or perhaps a Bald Eagle. Others may generally like wildlife but don't really pay attention to it until it lands, literally outside their window. When this happens, they may not know the name of the animal but they do recognize it as something they have not seen before. That is typically when a memory of watching a news story on some unusual city wildlife such as Peregrine Falcons pops into mind.

Still, not everyone who thinks he or she sees a Peregrine actually does. This chapter discusses a few species most commonly mistaken for Peregrines in Illinois. Along with physical characteristics, people should keep in mind the location of their observation as well as any behavior of the bird when trying to identify the species.

immature Pereg
Tail comparisons.
Adult Peregrine

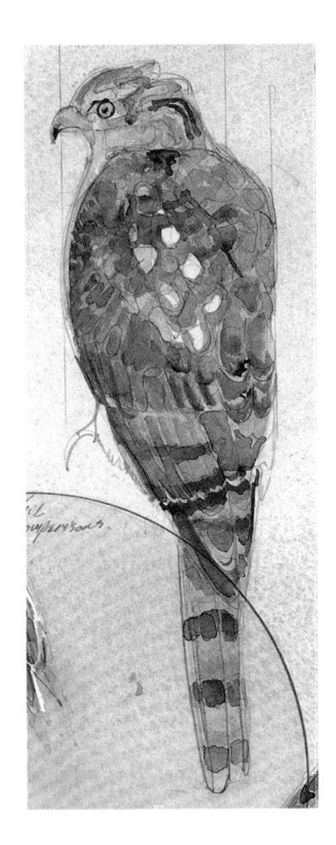

Cooper's Hawk
Accipiter cooperii
Order: Accipitriformes
Family: Accipitridae

Cooper's Hawks were once scarce in Illinois. Like the Peregrine, the hawks' adaptability to the urban environment has aided in their recovery. They are often sighted in neighborhood backyards feeding on birds at feeders. Old squirrels' or crows' nests may be reclaimed by the Cooper's Hawk.

Lack of a malar stripe, wide tail bands, and shorter wings distinguish Cooper's Hawks from Peregrines. Like Peregrines, immature Cooper's Hawks are brown with vertical streaks in the front. Adults are slate gray with barring on their front.

Red-tailed Hawk
Buteo jamaicensis
Order: Accipitriformes
Family: Accipitridae

Red-tailed hawks, the most common large hawk in North America, can often be seen along roadways. Plumage coloration is extremely variable, from light (Kriders) to almost black (Harlans). You'll find them atop telephone poles, eyes fixed on the ground to catch any movements of a vole, snake, or rabbit. Their attack is a slow controlled jump to the ground with legs outstretched—much different from a Peregrine's stoop.

To identify Red-tailed Hawks, look for the Buteo shape of broad, rounded wings with a short tail and field marks like the dark streaks along the abdomen (belly band). If it is in flight, check for marks on the leading edge of the wing. Only adults will have a red tail (immatures have banded tails).

American Kestrel

Falco sparverius

Order: Falconiformes

Family: Falconidae

Kestrels are the smallest falcons found in North America and the closest relative to Peregrines in Illinois. Data from migration studies, bird counts, and breeding bird surveys have all indicated a decline in kestrel populations over the past few decades. The cause of the decline is yet unknown.

Their small size and double malar stripe separate kestrels (7–8 inches) from Peregrines (13–23 inches.) Also, kestrels are cavity nesters while Peregrines prefer a "cliff" ledge.

Common Nighthawk
Chordeiles minor
Order: Caprimulgiformes
Family: Caprimulgidae

In urban environments, nighthawks nest on gravel rooftops making a scrape (depression) in the small rocks in which to lay their eggs. Like Peregrines, they do their hunting while in flight, taking advantage of clouds of insects including those attracted to streetlights.

Nighthawks have a similar shape to that of a falcon. Long tapered wings and a “hawk-like” face add to the confusion. Small size and white wing bars help to distinguish them from Peregrines.

CHAPTER TWENTY-FOUR

Urban Green Space

ARTIST NOTE: *I find it challenging to juxtapose a flat map with the areal perspective of a landscape painting. In one sense you are creating conflict, but the tension between the two gives life to the piece. Two of John Ruskin's laws of composition are contrast and interchange. Both laws "enforce the unity of opposite things." So both the flat map and landscape describe space in opposite ways and thus reinforce one another.*

Green space is vitally important for the health of both humans and city wildlife. Each component of its architectural design serves two purposes, functioning both as a space for wildlife as well as for human use and recreation. As cities grow, it has become important to set aside these green spaces where urban dwellers can enjoy trees, flowers, rivers, the lakefront, and wildlife.

Green spaces take many forms such as city parks, cemeteries, golf courses, bike and walking trails, beaches, and even plantings on rooftops. Landscape architects work with urban planners to integrate nature into an overall urban plan. It can sometimes be a challenge for humans and wildlife to coexist in the same habitat.

Montrose Dog Beach.
Montrose Beach
Montrose Point Bird Sanctuary
Harbor office
Magic Hedge
golf course
IMPORTANT BIRD AREA
SANCTUARY
NATURE

For birds, one of Chicago's most important areas is found at Montrose Harbor. Within 4.5 square miles, on a peninsula going into Lake Michigan, lay a beach, park, fishing pier, boating harbor, and grassland area. Adjacent to the south is a nine-hole golf course. The landscape that comprises Montrose serves as an important habitat for migratory birds. In order to protect the birds, a portion of the beach and the grassland area have been preserved as a bird sanctuary.

Bird sanctuary at Montrose Harbor

Because of the open space and presence of birds, Montrose attracts Chicago's resident Peregrines as a hunting ground, especially during the nonbreeding season. Similar areas along the lakefront (e.g., Northwestern Beach, Northerly Island, etc.) are vitally important green spaces within the city limits.

Montrose is an excellent place for bird watchers during migration. Its location and diverse habitats have made it a prime stopover location for passing birds.

Montrose Beach was a favorite location for the Uptown Theater Peregrines for many years. It is still used by Peregrines on a regular basis, as a hunting location and bathing spot, and at other times just as a nice place to roost.

Peregrines at the beach

Shorebird at Montrose Beach
Photo: J. Engel, 2012, Montrose Harbor, Chicago

Ruddy Turnstone
Arenaria interpres
Order: Charadriiformes
Family: Scolopacidae

Turnstones get their name from the manner in which they forage. Using their bills, they overturn stones, revealing the insects and aquatic invertebrates they feed upon.

LeConte's Sparrow

LeConte's Sparrow
Ammodramus leconteii
Order: Passeriiformes
Family: Emberizidae

LeConte's Sparrow is a ground forager, feeding on small seeds and arthropods. This secretive bird of the grasslands migrates through the Chicago area in spring and fall. It is seen annually at Montrose. The single highest count for these sparrows (according to ebird reports) was fifteen at Horseshoe Lake in the fall of 2008.

As of 2015, at least 339 bird species have been recorded at Montrose, including a number of rarities. The Montrose Sanctuary is a vitally important stopover location for migratory birds. With so many birds passing through, Montrose Harbor has seen its share of struggles. Whether it is dogs off their leashes, waterfowl trapped by frozen water, or, as in 2012, an outbreak of botulism, birds have faced their share of problems in addition to regular hardships of migration. Still, it is quite a benefit for migrating wildlife, if even for a short stopover, to have Montrose preserved.

Birds can either migrate in mixed company or keep to themselves. Radar is commonly used to track migrant flocks, allowing scientists and birders alike to study migration direction, density, altitude and how they relate to environmental factors.

Common Yellowthroat

Common Yellowthroat
Geothlypis trichas
Order: Passeriformes
Family: Parulidae

The Common Yellowthroat is just one of the small woodland warblers that pass through Chicago during spring and fall migration. Interestingly, the Common Yellowthroat was one of the first New World species described by Linnaeus in 1776.

Mixed flocks

Caspian Tern (center and right)
Hydroprogne caspia
Order: Charadriiformes
Family: Laridae

Ring-billed Gull (left)
Larus delawarensis
Order: Charadriiformes
Family: Laridae

The beach at Montrose is attractive to migrant shorebirds as well as wintering gulls. The Ring-billed Gull is one species that lives in Chicago year-round. Caspian Terns are seen during migration.

CHAPTER TWENTY-FIVE

Bird-Friendly Architecture

ARTIST NOTE: *Over the past twenty years, Chicago's government, architects, horticulturalists, and inhabitants have been building roof gardens as well as extending open spaces and parks. By 2015, Chicago was home to more than five hundred green roofs. The city's architectural history boasts of nature-based artists like Louis Sullivan and Frank Lloyd Wright. And today the work of Jeanne Gang adds to this legacy. Her Aqua tower and ecological designs like the Nature Boardwalk at the Lincoln Park Zoo go even further in cooperating with nature.*

Some birds live in Chicago year-round. Others for only a season. Yet others still, for only a few days, stopping to rest and refuel while passing through the city during migration. These are birds that are coming from their breeding ground that may be along the shore of the Arctic Ocean (and they may be on their way to extreme southern South America.) Migrant birds of all kinds may face population declines due to loss of breeding sites in the forests, wetlands and grasslands where they nest and in the tropical landscapes where they winter. Yet the urban landscapes they have to pass through enroute to their breeding and wintering grounds can be equally dangerous.

It is estimated that billions of migrating birds die each year from striking building windows. More die from hitting wind turbines or TV towers. For raptors, wind turbines are more dangerous, as they are erected in the open spaces raptors favor for hunting and pass through when migrating. They generate winds, which ease flying. Scavengers, such as eagles, are attracted to birds maimed or killed by the turbines.

Window kills are more commonly nocturnal migrants, smaller-sized species that fly through the city at night. Some casualties happen during the day when glass windows reflect the landscape, making a bird think it can land in the image

McCormick Place
Hancock
Roosevelt University

of a tree shown on the glass. Window kills for Peregrines are likely to occur when a Peregrine is chasing prey and either doesn't see the window, or can't get out of the way in time.

What can be done to help reduce the loss of migrant birds? Architects, building managers, engineers, and concerned citizens have banned together with scientists to find ways of improving avian safety. For instance, the design of wind turbines has changed over time to minimize surface area of blades. Fewer turbines placed strategically out of major migration routes can reduce the number of birds adversely affected. In cities, architects can design a building using materials like bird-safe nonreflective glass that help lower the chances of migrant strikes. Groups of individuals, such as the Chicago Bird Collision Monitors, have rescued thousands of stunned birds and taught building managers that bird loss can be reduced by simply turning off the lights.

There are a number of ways that architects can alter a city skyscraper to make it friendlier for birds. Some window collisions occur during daylight and are caused by the reflection of landscaping in the glass. Birds fail to see the glass. Bird-friendly glass was developed with a patterned UV-reflective coating, revealing it to birds while remaining virtually invisible to the human eye.

Chicago's nighttime skyline is darker these days. Why? Lights can be problematic for avian migrants. It creates a false star pattern that confuses birds. In stormy weather, buildings with large bright single lights, like the John Hancock, can be the single beacon guiding the bird, which fails to see the structure hidden in the clouds. Studies at the Field Museum have shown that by closing curtains or turning off the lights, the number of window kills can be reduced by over 80 percent.

Building design

Lights

White-throated Sparrow

White-throated Sparrow
Zonotrichia albicollis
Order: Passeriformes
Family: Emberizidae

White-throated Sparrows are a common migrant in Chicago. They can be found wintering in Illinois, and they breed in the boreal forests of Canada and the Northeastern United States.

Yellow Rail

Yellow Rail
Coturnicops noveboracensis
Order: Gruiformes
Family: Rallidae

Yellow rails are a rare migrant in Illinois and a challenging find for most birders. This secretive marsh bird frequents wet grassy areas, meadows, and hayfields.

The Field Museum's Chicago Bird Salvage Dataset is the result of window-casualty collecting in Chicago. From a single building studied since the late 1970s and the city's downtown Loop since 2002, over 72,000 individual migratory birds have been collected to date. This database has been used to examine trends in migration as well as document the rare, such as a single Yellow Rail specimen found in the fall of 2010.

LANDSCAPING

Landscaping can be a problem for migrant birds when it gets reflected in glass surfaces. It can also be a tool to improve survival if it blocks the glass or be a place for birds to stop, rest, and refuel. At a building that ornithologists at the Field Museum have studied for thirty-five years, the landscaping grew to the point where it created a buffer between the building itself and the surrounding green space, where migrant birds would land.

GREEN ROOFS

In Chicago. . . .

In 2015, nearly seven million square feet of greenery can be found on over five hundred rooftops. They provide stop-over sites for tired migrants. In addition, some roofs grow produce for restaurants or serve as classrooms for urban agriculture.

Green roofs can be cost effective; they extend roof life and reduce energy expenses.

CHAPTER TWENTY-SIX

Conservation and Natural History Museums

ARTIST NOTE: *I have enjoyed the endless wonders of museum collections for the past thirty-five years. While they afford the scientist documentation and study opportunities, they teach the artist as well. They expose endless design displays and color comparisons. There is magic in how things of nature are put together. The artist appropriates this magic by osmosis.*

Why a chapter on conservation and natural history museums in a book about Peregrines? Natural history collections are a baseline of information used to understand environmental concerns of the day. The recovery of Peregrine Falcons is a prime example of the importance of such collections. Eggs collected years prior to the decline of the species were vital to discovering the connection to pesticide contaminants. This understanding became the first step in the recovery of the species.

Every natural history specimen that has good data, such as the date and location where it was found, is a record of that species (or community) in time and space. That individual specimen can be compared to others found in different localities. Or perhaps the item will be studied against the same type of specimen found a hundred years ago. Photographs alone cannot provide the detailed information that the actual specimen can.

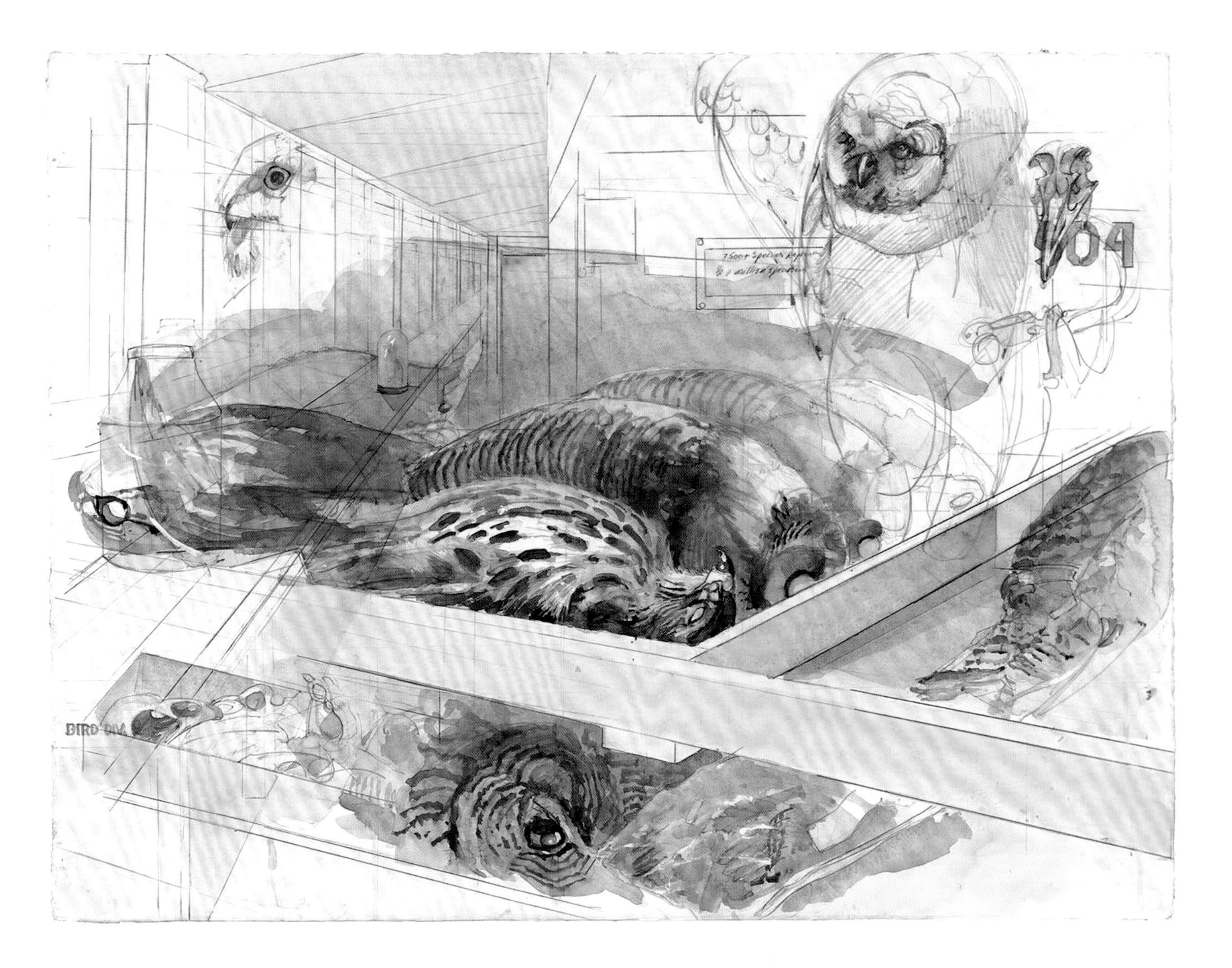
04
BIRD DIV

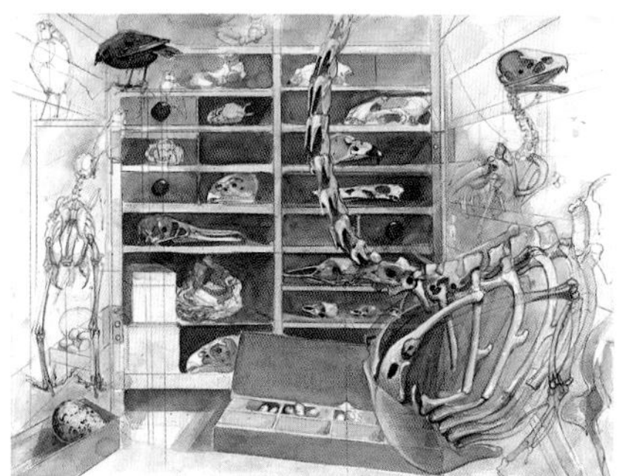

EXHIBITS

Exhibits engage and inspire visitors to a museum to discover more about their world. Whether focused on natural history or having an anthropological theme, an exhibit display creates a glimpse for the public into something they may never experience in life.

NATURAL HISTORY COLLECTIONS

The breadth of preservation and use surrounding a single individual specimen within a natural history museum can be astounding. By just coming in the door, that specimen (let's say a bird) can expand our knowledge about its life in the natural world. Scientists record where and when the bird was obtained. Ectoparasites (microrganisms living on the carcass such as mites and lice) may be collected. Any information that will be lost in preparation into a museum specimen is recorded—weight, body measurements such as wing chord and bill length, fat levels, and so on. Gonads are measured to verify sex, and tissue is taken for later studies on genetics. For birds, a choice is made to keep the specimen as either a complete specimen preserved in alcohol, a study skin, or a skeleton.

Once catalogued into the collection, the specimen itself is available for use. The information accompanying it may have been studied at any time. The bird collection at the Field Museum has been used by a wide variety of individuals: from researchers to teachers, artists, ecologists, photographers, students, and generally anyone with an interest in the avian world.

Taken as a whole, museum collections and exhibition materials represent the world's natural and cultural wealth. As a steward of that wealth, the Field Museum advances an understanding of all natural forms and of the human experience. This institution serves as a resource for humankind, and in all its activities fosters an informed appreciation and understanding of our rich and diverse world. It is important to preserve that inheritance for future generations. The museum's Gidwitz Hall of Birds and Abbott Hall of Conservation include a mounted Peregrine to tell the story of the species' decline and recovery.

The Field Museum's Integrative Research and Collections Centers are at the leading edge of scientific research, with more than 140 scientists conducting research in our labs and around the globe to answer key questions about our world and heritage. They rely on the museum's twenty-six million artifacts and specimens—only 1 percent of which are on public display—to answer fundamental questions about our planet, its life, and its cultures.

CHAPTER TWENTY-SEVEN

A Species Recovered

SCIENTIST NOTE: *The very first Peregrine I ever saw was Harriet, the adult female at Chicago's Wacker site. She was stooping on a Yellow-bellied Sapsucker, zipping with great speed around neighboring skyscrapers. It was amazing; I was hooked after my first sight of that bird. I had no idea at the time that I would spend the next thirty-plus years working with the species. Or imagine that I would have the opportunity to band over five hundred Peregrine nestlings. I didn't expect that the population would grow to encompass the entire state within such a short period of time. While those are things I couldn't foresee, I have always known it was a privilege to do this work. I am grateful to have played a small part in the recovery of Peregrines.*

Peregrines in Illinois are doing great! Illinois officially removed Peregrine Falcons from the state endangered and threatened species list in May 2015.

What does this mean for our state population of Peregrine Falcons? First, it is an affirmation of the recovery of the species in Illinois. This is due in large part to the long-term stewardship and dedicated effort of the numerous individuals and organizations that have supported and assisted in looking after Illinois' Peregrines.

It does not leave our Peregrines unprotected. The Migratory Bird Treaty Act makes it "illegal for anyone to take, possess, import, export, transport, sell, purchase, barter, or offer for sale, purchase, or barter, any migratory bird, or the parts, nests, or eggs of such a bird except under the terms of a valid permit issued pursuant to Federal regulations." Peregrines are protected under this law.

Another thing the Peregrine's removal from Illinois' endangered and threatened species list will not change is the work of the Chicago Peregrine Program. The program oversees Illinois' Peregrine Falcon population. This is accomplished by monitoring individual birds and nest sites, conducting scientific research, and participating in public education.

Historic eyries were along the Illinois and Mississippi Rivers. Interestingly, an 1889 record reported by ornithologist Robert Ridgeway recorded Illinois Peregrines nesting in the hollow branches of very large sycamore tree! While we don't currently have any tree-dwelling Peregrines in the state, we have seen Peregrines return to natural and artificial cliffs (quarries) as well as thrive in an urban habitat.

The goal to reestablish Peregrines breeding in Illinois was met with our urban falcons. We had hoped we would eventually see a return to natural cliff sites. This was realized when Peregrines began nesting in the Alton area of southern Illinois.

Though Peregrines have returned to Illinois' natural cliffs, by far the greatest population growth has occurred in the Chicago area. When Peregrines first returned to Illinois, the average territory size by a single bird was five square miles. As the population has grown, the size of the territory has decreased. As of 2015, six pairs of Peregrines were nesting in one square mile of real estate in downtown Chicago.

Illinois cliff sites

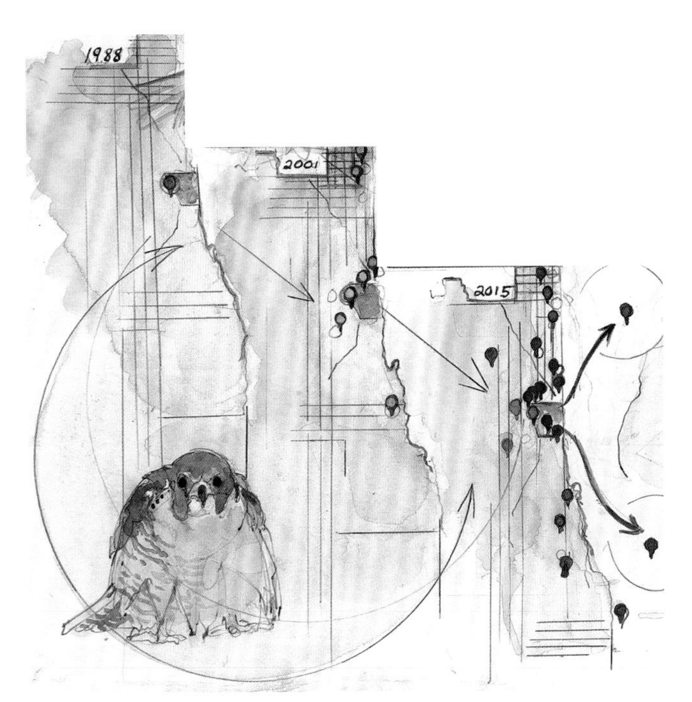

Peregrine population growth

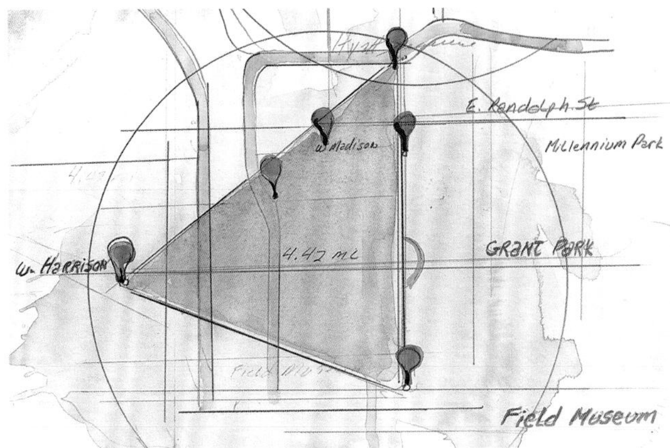

2015 Peregrine nest cluster

Illinois' first post-decline breeder: Harriet
Photo: M. Hennen, 1990

Harriet's Productivity: 1987–97
Total Eggs: 37
Eggs hatched: 26
Young fledged: 24

The first Peregrine egg laid in Illinois after the decline was at the Wacker site in Chicago. In 1987, Harriet, a Minnesota-hacked bird, met up with Jingles, a 1985 Chicago release. An infertile egg was laid on a building near Chicago's famed Sears Tower. The following year, Harriet and Jingles raised two young, the first Peregrines born in the state since 1951. While Jingles was at the site for only a few years, Harriet continued to breed at Wacker for the next eleven years until she was injured.

The Wacker site, Illinois' first post-decline breeding location and longest-running site, remains the hallmark of Peregrine recovery for our state.

WHAT NOW?

With the species doing well, people working in Peregrine recovery must decide where to go to from here. I don't know that I have the answer and certainly each individual or group must decide what is best for their situation. Some challenges will be the same for all states. How do we to keep stakeholders (building owners and landowners) engaged and accepting of the species? Can we foster conservation values in the public?

How do we reach future generations to tell the Peregrine's story?

In the Peregrine world, the work of individuals and groups like the Chicago Peregrine Program has many facets. We are scientists, educators, researchers, and public relations managers all in one. We document what is occurring with the species while working to best assist the species to continue to breed. I don't see that changing.

With such a high percentage of the Peregrine population being urban, I think it becomes ever more important for someone to continue to teach the public about the species and how to cope with living in close proximity to them. We must remain a liaison for the Peregrine to the public. I believe the second-greatest need is education. To reach younger generations it is prudent to keep up with the times—even using social media as a way to talk about the species. Better yet, we should use the Peregrine story as a springboard to other conservation issues such as habitat change and global warming. If everyone does a small amount of work towards conserving the environment, collectively we can make a great difference.

The Chicago Peregrine Program is a wide and diverse group, all volunteers with a united interest and concern for Peregrines. I believe that has been the key to our success; whatever we do, we must find a way to continue our role as Peregrine ambassadors.

ACKNOWLEDGMENTS

Since its inception, the Chicago Peregrine Program has been a wide network of volunteers, all with a goal of helping reestablish Peregrines back in the wild. From Peregrine watchers to building owners, engineers, management personnel, wildlife rehabilitators, vet technicians, city workers, collision monitors, office workers, condo residents, librarians, power plant personnel, photographers, hack site attendants, technicians, educators, and everyone who has every let us know of a Peregrine sighting—first and foremost—all of you deserve a round of applause.

We would have loved to highlight each individual Illinois Peregrine and every site whether active or not. This was not possible. My apologies if your favorite bird or site is not mentioned in the text.

Many thanks to Stephanie Ware, Josh Engel, and Matthew Gies, all close friends and the core of the Peregrine team for many years. Words cannot express my gratitude and friendship towards you.

Thanks to George and Berni Richter (SOAR—Save Our American Raptors) and the many folks at Willowbrook Wildlife Center for your fine care of our injured Peregrines. Thanks to Joel Pond, Deborah Cohen, Virgil Silva, Dave Syfczak, Isabel Caballero, David Mendez, Sean Ware, and Kristopher Lah.

Our gratitude to all who have been personnel or volun-

teers of the Chicago Peregrine Program. To the Raptor Center, state coordinators, the Midwest Peregrine Society, and the city of Chicago. Thanks!

Thanks to everyone who provided input on this book, most importantly, John Bates. Also, thanks to my family and my museum colleagues.

Finally, my thanks to the late Vicki Byre, my mentor and friend, who opened my door to the world of Peregrines.